INVASIVE LIONFISH:
A Guide to Control and Management

Editor
James A. Morris, Jr.
National Oceanic and Atmospheric Administration

Format & Graphics
LeRoy Creswell
Florida Sea Grant
Gulf and Caribbean Fisheries Institute, Inc.

Publishing Consultant
Ann Boese
Laughing Gull, Inc.
Savannah, Georgia, United States

Gulf and Caribbean Fisheries Institute
Special Publication Series, Number 1

GULF AND CARIBBEAN FISHERIES INSTITUTE, INC.

Marathon, Florida, United States

GULF AND CARIBBEAN FISHERIES INSTITUTE, INC.

This Publication Is Available From:

Gulf and Caribbean Fisheries Institute, Inc.
2796 Overseas Highway, Suite 119
Marathon, Florida 33050 USA
http://www.gcfi.org
or
NOAA Central Library
http://www2.coastalscience.noaa.gov/publications/

Suggested Citation Formats

Entire Book

Morris, J.A., Jr. (Ed.). 2012. *Invasive Lionfish: A Guide to Control and Management.* Gulf and Caribbean Fisheries Institute Special Publication Series Number 1, Marathon, Florida, USA. 113 pp.

Chapters Within the Book

Akins, J.L. 2012. Education and outreach: Building support and expertise. Pages 15-23 in: J.A. Morris Jr. (ed.) *Invasive Lionfish: A Guide to Control and Management.* Gulf and Caribbean Fisheries Institute Special Publication Series Number 1, Marathon, Florida, USA. 113 pp.

ISBN: 978-0-615-69974-5

Library of Congress Catalog Number 2012949444

CONTENTS

AUTHORS

James A. Morris, Jr. (Editor),
Ecologist
National Centers for Coastal Ocean Science
National Ocean Service, NOAA
Beaufort, North Carolina, United States
james.morris@noaa.gov

J. Lad Akins
Director of Special Projects
Reef Environmental Education Foundation
Key Largo, Florida, United States
lad@reef.org

Dayne St. A. Buddo
Marine Ecologist, Lecturer, and Academic Coordinator
Discovery Bay Marine Laboratory
University of the West Indies, Jamaica
dayne.buddo@uwimona.edu.jm

Stephanie J. Green
Simon Fraser University
Burnaby, British Columbia, Canada
steph.j.green@gmail.com

Ricardo Gómez Lozano
Director, Parque Nacional Arrecifes de Cozumel
Comisión Nacional de Áreas Naturales Protegidas de México
Cozumel, México
rglozano@conanp.gob.mx

CONTRIBUTORS

The strategies and practices provided herein were developed during the 2010 International Coral Reef Initiative–Caribbean Regional Lionfish Workshop in Cancun, with participants from the following organizations:

Comisión Nacional de Áreas Naturales Protegidas, México
University of the West Indies, Discovery Bay Marine Laboratory, Jamaica
Department of Natural and Environmental Resources, Puerto Rico, United States
U.S. Department of State
U.S. National Oceanic and Atmospheric Administration
Marine Protected Areas, Belize
Department of Marine Resources, Bahamas
Department of Environment, Cayman Islands
Bermuda Aquarium, Museum, and Zoo
CSI-ICRI Coral Reef Enforcement and Investigation
Environment State Service, Martinique, French West Indies
Specially Protected Areas and Wildlife–Regional Activity Center
The Nature Conservancy
National Environment Centre, Turks and Caicos
CORALINA, Columbia
Centre for Agricultural Biosciences International
Acuariología, Acuario Nacional, Cuba
Universidad Autónoma de Nuevo León, Mexico
Nature Foundation, St. Maarten, The Netherlands
Ictiología, Museo de Historia Natural, Universidad La Salle, Venezuela
Bonaire National Marine Park, Bonaire, Caribbean Netherlands
Reef Check Dominican Republic
Reef Environmental Education Foundation
International Coral Reef Initiative
Seatone Consultants

FOREWORD

From the International Coral Reef Initiative

The proliferation of lionfish in the Caribbean over the last ten years is a real and growing threat to the ecology of tropical and sub-tropical marine areas in that part of the world. Addressing this issue requires a collaborative approach at a regional, national, and local scale.

The International Coral Reef Initiative (ICRI), as an informal collective, is well placed to provide a mechanism for cooperation on the lionfish issue. The facilitation of the Caribbean Regional Lionfish Workshop with key Caribbean players in 2010, and the establishment of an ad-hoc Committee on Caribbean Regional Response to the Lionfish Invasion later that year — now referred to as the Regional Lionfish Committee — are practical examples of the support that ICRI has provided to assist in addressing the lionfish issue.

Invasive Lionfish: A Guide to Control and Management is an important step on an urgent and challenging journey. The International Coral Reef Initiative is happy to be involved and will continue to support important actions, such as the publication of this guide.

Margaret Johnson	***Beverly Wade***
Co-chair, ICRI Secretariat	***Co-chair, ICRI Secretariat***
Great Barrier Reef Marine Park Authority	***Belize Fisheries Department***
Australia	***Belize***

ICRI

INTERNATIONAL CORAL REEF INITIATIVE

The International Coral Reef Initiative (ICRI) is an informal partnership of like-minded governments and organizations seeking to build and sustain the capacity of countries and regions to achieve effective management and sustainable use of coral reefs and related environments. While the ICRI supports actions from the Regional Lionfish Committee, including the publication of this manual, the ICRI makes no representation about the content and suitability of this information for any purpose. Inclusion in this document of any specific action doesn't represent a specific endorsement from ICRI of that strategy.

From UNEP–Caribbean Environment Programme

The Convention for the Protection and Development of the Marine Environment in the Wider Caribbean Region (WCR) (Cartagena Convention) is the only legally binding regional environmental treaty for the Region, and together with its supporting Protocols, constitute a legal commitment by these countries to protect and manage their common coastal and marine resources individually, jointly, and in a sustainable manner. One out of the three protocols of the Cartagena Convention, the Protocol on Specially Protected Areas and Wildlife (SPAW) was adopted in 1990 and entered into force in 2000. Sixteen countries are currently Parties to the Protocol, the newest ones being Guyana in 2010 and Grenada and The Bahamas in 2012. UNEP's Regional Coordinating Unit for the Caribbean serves as Secretariat to the Convention and its Protocols, and its Caribbean Environment Programme (CEP) provides the technical support to Governments to meet their obligations.

The objectives of the SPAW Protocol are to protect, preserve, and manage in a sustainable way: 1) areas and ecosystems that require protection to safeguard their special value, 2) threatened or endangered species of flora and fauna and their habitats, and 3) species, with the objective of preventing them from becoming endangered or threatened. The SPAW Protocol stresses the importance of protecting habitats as an effective method of protecting the species. Protection is focused on fragile and vulnerable ecosystems as a whole, rather than on individual species. In that context, invasive alien species are of particular concern for the SPAW Protocol and its Programme of Work. The current SPAW Workplan, adopted in October 2010 by the 6th Conference of the Parties to the SPAW Protocol, includes support to regional initiatives for prevention and mitigation of the impacts of invasive species, in particular to the lionfish.

UNEP-CEP and its SPAW Regional Activity Center (SPAW-RAC), hosted by the Government of France in Guadeloupe, have in that context joined efforts with ICRI and numerous partners to address the lionfish issue in the Caribbean. We are grateful for the efforts and dedication of the writing team, of its leader James Morris, and of all the authors that devoted time and skills for the preparation of the various chapters that compose the manual, *Invasive Lionfish: A Guide to Control and Management* is now available. We believe this guide will prove to be a crucial tool for managers and practitioners in the Wider Caribbean Region, and we hope it will help them better control the lionfish invasion and its impacts by providing practical solutions and recommendations. The translation of the guide into Spanish, supported by the SPAW -RAC, will facilitate its use by practitioners in all Caribbean countries and territories.

We look forward to the wide use of the *Invasive Lionfish* guide in all countries of the region and to the successful control of this marine invasive species.

Nelson Andrade Colmenares, Coordinator
UNEP–Caribbean Environment Programme

UNEP

PREFACE

The countries of the Wider Caribbean with their vibrant coral reefs are major tourist destinations, bringing billions of dollars to the Caribbean annually. The importance of reef resources is invaluable to local economies and cultures. Invasive Alien Species (IAS) are considered one of the major threats to native species and habitats in the Caribbean. Recognizing the urgency of the lionfish invasion and the need to develop clear recommendations for local control, a special workshop was organized by an international team, including the International Coral Reef Initiative, REEF, NOAA, and SPAW-RAC, and it took place in August 2010 in Cancun, Mexico. With 47 participants representing over 25 organizations from 20 countries and territories of the Wider Caribbean, the ICRI Regional Lionfish Workshop identified many of the best strategies for addressing the lionfish invasion.

The intent of this publication, *Invasive Lionfish: A Guide to Control and Management*, is to provide a reference for resource managers, policy makers, field workers, outreach coordinators, researchers, fishers, divers, and the general public — who are actively engaged in learning about lionfish and developing local control strategies. The strategies recommended herein are based upon best available science and practice, which in time will change as new approaches and the effectiveness levels of existing approaches emerge. Resource managers are encouraged to use the information in this guide to develop local management plans.

James A. Morris, Jr.
National Centers for Coastal Ocean Science, NOAA

ACKNOWLEDGMENTS

This publication, *Invasive Lionfish: A Guide to Control and Management*, would not have been possible without the dedicated support of the many contributors and project team members, who assisted in the early conceptualization of the strategies provided herein. The authors are especially grateful to Christy Loper (NOAA Coral Reef Conservation Program), Catherine Toline (U.S. National Park Service), Isabelle Côté (Simon Fraser University), and Mark Albins (Oregon State University) for their insights into the monitoring chapter.

The authors also thank the many reviewers of the original Cancun Lionfish Working Group for their helpful comments, and Patti Marraro (NOAA), Jeff Govoni (NOAA), Paula Whitfield (NOAA), Pam Schofield (USGS), Nancy Brown-Peterson (University of Southern Mississippi), and David Johnson (NOAA), whose comments also greatly improved this manuscript.

We especially thank Dr. Mark Hixon and his doctoral students at Oregon State University, all of whom are studying the lionfish invasion, including Mark Albins, Cassandra Benkwitt, Kurt Ingeman, Tye Kindinger, Timothy Pusack, and Lillian Tuttle, for their careful review. Lastly, we thank Nancy Daves and the NOAA National Marine Fisheries Service Office of International Affairs for providing the funding for this work. — *James A. Morris, Jr.*

Project team members pictured with Alfredo Arellano Guillermo, Director General de Conservacion para el Desarrollo CONANP, Mexico, who called for implementation of the lionfish management strategies provided in this guide at the 64th Gulf and Caribbean Fisheries Institute in Puerto Morelos, Mexico, October 31–November 4, 2011. (*From left to right*) Scot Frew (NOAA), Dayne St. A. Buddo (University of the West Indies), J. Lad Akins (REEF), James A. Morris, Jr. (NOAA), Alfredo Arellano Guillermo (CONANP), Nancy Daves (NOAA), Ricardo Gómez Lozano (CONANP), and Franck Gourdin (SPAW-RAC).

ACRONYMS

AGRRA	Atlantic and Gulf Rapid Reef Assessment
CABI	Centre for Agricultural Bioscience International
CITES	Convention on International Trade in Endangered Species of Wild Flora and Fauna
COB-MESI	The College of the Bahamas Marine and Environmental Institute
CONANP	Comisión Nacional de Áreas Naturales Protegidas
CORAL	Coral Reef Alliance
CORE	Caribbean Oceanic and Restoration and Education Foundation
E&O	Education and Outreach
FKNMS	Florida Keys National Marine Sanctuary
IAS	Invasive Alien Species
ICRI	International Coral Reef Initiative
IMCC	International Marine Conservation Congress
ISSG	Invasive Species Specialist Group
MTIASIC	Mitigating the Threats of Invasive Alien Species in the Insular Caribbean
NLRP	National Lionfish Response Plan
NOAA	National Oceanic and Atmospheric Administration
NOS	National Ocean Service
REEF	Reef Environmental Education Foundation
SPAW-RAC	Specially Protected Areas and Wildlife–Regional Activity Center
STINAPA	Stichting Nationale Parken (Bonaire Marine Park)
UNEP	United Nations Environment Programme
UPRM	University of Puerto Rico–at Mayagüez
USGS	United States Geological Survey

CHAPTER 1

THE LIONFISH INVASION: *Past, Present, and Future*

James A. Morris, Jr.

The invasion of lionfish (*Pterois miles* and *P. volitans*) may prove to be one of the greatest threats of this century to warm temperate and tropical Atlantic reefs and associated habitats. As the first marine reef fish invasive species to this region, lionfish are changing the culture of how reef managers view invasive species, the regional connectivity of marine reefs, and their vulnerability to marine invasions.

The term "lionfish" is now as notorious as the other major invaders of the last century, such as Asian carp, kudzu, zebra mussels, and sea lamprey. Originally imported into the United States as a popular aquarium fish, the lionfish is now one of the most abundant top-level predators of many reefs. Lionfish pose a threat to the integrity of the reef food web and are capable of impacting commercial fisheries, tourism, and overall coral reef health.

Viewed in context with other reef stressors — such as land-based pollution, climate change, and overfishing — the lionfish invasion is distinguished by two obvious characteristics.

A RAPID, WIDESPREAD INVASION

The first outstanding characteristic of the lionfish invasion is that it has occurred rapidly across a wide geographic area. The initial confirmed lionfish sighting in the United States occurred in 1985, off Dania Beach, Florida. Some 15 years later, in 2000–2001, lionfish were identified as an established invader in the offshore waters of North Carolina, United States. At the time of this writing in 2012, lionfish are fully established throughout the Southeast United States, the Caribbean Sea, and much of the Gulf of Mexico. Lionfish are expected to reach the warm temperate reefs of South America soon.

BROAD DIET AND GENERAL HABITAT

The second pronounced characteristic of lionfish is their broad diet and general habitat preferences. Because of these factors, lionfish have the potential to affect the structure and function of many Atlantic marine communities — from the sea surface to depths exceeding 300 meters, and across habitats ranging from coral and hardbottom to artificial reefs, mangroves, and seagrass beds. For example, the high densities of lionfish observed in locations such as the Bahamas may be causing an abrupt change to the biodiversity and community structure of reef fish communities, and could constitute the most significant change since the beginning of industrialized fishing (Albins and Hixon 2011).

CASCADING IMPACTS

Alarmingly, lionfish may trigger cascading impacts through their disruption of the food web. For example, the lionfish consumption of herbivorous fishes could reduce the functional role of herbivores in keeping algae in check, a process known to be important for the health of coral reefs. Lionfish may also compete for resources – principally food and space – with economically important species, such as snapper (Lutjanids) and grouper (Epinephelids). It is uncertain if stock-rebuilding efforts will be able to return reef fish stocks to pre-lionfish abundance levels.

Lionfish could also affect the recovery of species of concern, such as the Nassau grouper (*Epinephelus striatus*), Warsaw grouper (*E. nigritus*), and speckled hind (*E. drummondhayi*). These species are critically low in abundance and might not recover quickly under the additional predation mortality imposed by lionfish.

Lastly, it is the interaction of the lionfish invasion with existing reef stressors that poses the greatest concern. Coral reefs of the Atlantic are already highly stressed from bleaching events, climate change, ocean acidification, overfishing, and pollution. The additional stress of this invasive species could accelerate and compound the degradation of coral reef ecosystem health in profound and unexpected ways.

LOCAL INTERVENTION IS CRITICAL

Local control efforts are critical for mitigating the effects of lionfish on key marine habitats. These efforts are invaluable for supporting other conservation initiatives, such as management of marine protected areas and fisheries stock rebuilding. The re-colonization of lionfish from remote and unmanaged habitats will continue to inflict constant stress on Atlantic marine communities. Until new technologies and approaches are developed for controlling lionfish populations, managers must be prepared for long-term intervention.

CHAPTER 2

LIONFISH RESEARCH:
Current Findings and Remaining Questions

James A. Morris, Jr. and Stephanie J. Green

Our understanding of lionfish biology and ecology has come primarily from scientific research activities. Both published literature and direct communications with researchers are included in this synopsis on the state of knowledge of lionfish distribution, biology, and ecology in their invaded and native ranges. Many new questions will emerge capable of teaching us not only about lionfish biology and ecology, but also about how Atlantic reef systems are responding to the invasion. As the invasion continues to intensify and spread to new locations, so does the demand for new research to inform management decisions.

Research is an investigation using the scientific method to generate new information about a subject or develop novel approaches to solve problems. Research findings can inform resource management by:

i) Providing the best possible information upon which to base decisions, and
ii) Uncovering new approaches to persistent problems.

It's important to note that research differs from monitoring in its purpose, though both use similar quantitative methods to collect information. Monitoring (see Chapter 5) is the process of collecting specific information that describes trends on the status of a subject for which there is already some knowledge, but for which management guidance is needed. Within research, however, the pursuit of information can be hypothesis driven or descriptive, and it ultimately results in learning something new about the subject matter itself. For example, lionfish stomach content analysis can provide new information on lionfish diet (i.e., research), but it can also be used to track changes in diet composition and impacts at a local scale (i.e., monitoring).

Research priorities for management will vary across space and time depending on the amount of resources available and the urgency of the information the research activity provides. Managers who support lionfish research in their local area increase the chances that new approaches will be developed to control lionfish.

RESEARCH FINDINGS TO DATE

Distribution and Densities

Geographic distribution

Sightings and collection reports indicate that lionfish are established in the offshore waters of the Southeast United States and throughout most of the Caribbean, and are presently invading the Gulf of Mexico and parts of South America (Figure 2.1) (Schofield 2009, 2010, Schofield et al. 2012). Lionfish are expected to invade the remainder of the Caribbean and to continue the southward expansion along the coast of South America until water temperatures below the lionfish thermal tolerance limit continued expansion (Figure 2.2) (Morris and Whitfield 2009). Lionfish have been reported from all major marine seafloor and substrate types within the invaded Atlantic, and they occupy a range of depths – from shoreline to >300 m depth – as recorded in the USGS Invasive Species Sightings Database. While the majority of research on the invasion has taken place on coral reefs and hardbottom habitats, lionfish also occupy mangrove (Barbour et al. 2010), estuary (Jud et al. 2011), seagrass, and artificial structures (Smith 2010).

Figure 2.1 Reports of lionfish as of June 2012.

Figure 2.2 Potential future distribution of lionfish, assuming sea surface temperature will be the only limiting factor.

Invasion status

The population growth of many invasive species follows a predictable trajectory, including a lag phase, followed by a period of exponential growth, an invasion peak, and then equilibrium. For many invasive species, population densities at the invasion peak can exceed the carrying capacity of the new system (Figure 2.3). Equilibrium is reached when populations of the invasive species are limited by factors such as competition with native species or themselves for food or space, predation (including cannibalism), parasitism and disease, and abiotic factors such as temperature.

Invasive lionfish populations across the Atlantic appear to be following this general course. The earliest sighting of lionfish in the Atlantic dates to 1985 off the southeastern coast of Florida (Morris and Akins 2009). Over the next 15 years, lionfish were observed sporadically along the Florida coast. In 2000, multiple individuals were sighted off North Carolina, South Carolina, and Georgia; and from 2000 to the present, lionfish densities have continued to increase rapidly (Morris and Whitfield 2009, Green et al. 2012a). Some locations have reported decreases in lionfish densities, though it is difficult to disentangle the effect of removal from natural processes in controlling lionfish abundance at a local scale. Given that lionfish have only recently established in the Gulf of Mexico and southern Caribbean, increases in abundance in these habitats are expected.

Long-term, systematic monitoring of lionfish densities is required to construct population growth curves for lionfish across the invaded region (see Chapter 3). The implementation of such programs early in the invasion stage will provide critical information to managers on the speed of the invasion and the effectiveness of control. This information will aid in allocating management resources over space and time (see Chapter 4). Rigorous indices of abundance, similar to those developed for monitoring native fishes, are also needed to assess annual changes in lionfish densities and to determine whether densities are increasing, decreasing, or remaining constant throughout the invaded range. Unfortunately, few monitoring efforts of this type are presently underway.

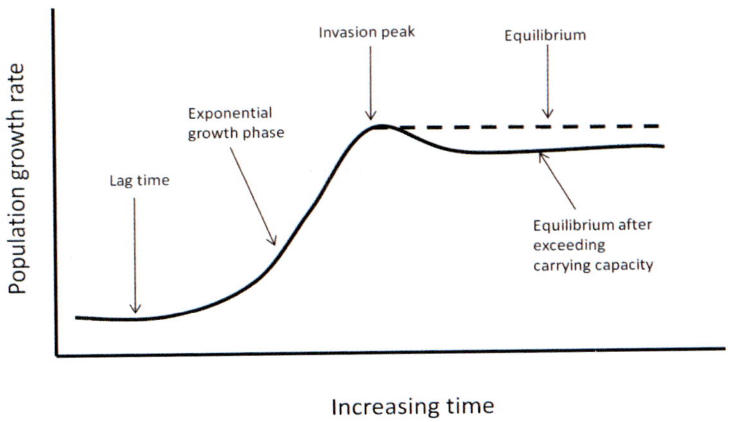

Figure 2.3 Theoretical line-plot depicting a simplified population growth curve for invasive species. Line depicts a smoothed curve as marine fish population dynamics can exhibit wide variation in annual recruitment.

Biology and Ecology

Taxonomy and genetics

Analysis of mitochondrial DNA indicates that two species, the devil firefish (*Pterois miles*) and the red lionfish (*P. volitans*), were introduced into the Atlantic (Hammer et al. 2007, Freshwater et al. 2009). These two species are identical morphologically in the Atlantic (Hammer et al. 2007), but in their native range they can be distinguished with meristics; *P. volitans* exhibit one higher count of dorsal and anal fin rays when compared to *P. miles* (Schultz 1986). Devil firefish is normally found in the Red Sea, Persian Gulf, and Indian Ocean (excluding Western Australia) and red lionfish is found in the Western and Central Pacific and Western Australia (Schultz 1986). In the Atlantic, *P. volitans* is documented throughout the Southeast United States and Caribbean with *P. miles* only documented along the Southeast United States and the Bahamas (J. Morris, Unpublished data).

As the lionfish invasion continues, genetic research will become a powerful tool for assessing various dispersion pathways, divergence of sub-populations, expression of phenotypes that are driving invasiveness, and possibly detection of new introduction events. Freshwater et al. (2009) provided the first such comparison and reported high similarity between lionfish captured in North Carolina and the Bahamas. This result, coupled with information on oceanographic current patterns, suggests that Bahamian lionfish may have originated from the East Coast of the United States. Subsequent study has revealed a secondary founder effect with the continued dispersal of lionfish into the Caribbean and also that the invasion's progression supports hypotheses of connectivity for tropical western Atlantic marine organisms (Betancur-R. et al. 2011). The development of high-resolution lionfish genetic markers will allow a more thorough investigation of this invasion and provide a better understanding of invasion genetics and the region's marine phylogeography.

Diet and foraging behavior

Through stomach contents analysis, Morris and Akins (2009) described lionfish as generalist carnivores that feed on a wide variety of fishes and crustaceans. Visual observations of lionfish on invaded reefs indicate that lionfish consume prey at high rates, largely during crepuscular periods (Green et al. 2011). Daily patterns of lionfish behavior appear to be consistent with reports from the native range, with the species most active and foraging out over habitat during low-light crepuscular periods during dawn and dusk (Green et al. 2011). Though consistent with lionfish behavior in the native Indo-Pacific, their hunting strategy is largely unique among fish predators in the Caribbean. Lionfish hover motionless over prey with their large pectoral fins extended and are able to approach their prey closely before making a rapid strike. Sometimes, lionfish extrude water jets to orient the prey towards the mouth before striking (Albins and Lyons 2012). Prey are typically consumed whole.

During the day, lionfish are often found resting and sheltered under ledges and overhangs, associated with large high-relief structures within the habitat (Green et al. 2011, Darling et al. 2011). Ambient light levels may significantly affect temporal patterns of lionfish behavior. Observations of lionfish under variable weather conditions found that activity levels and predation rates during the day were significantly higher during overcast times, when light levels may simulate those of crepuscular dawn and dusk periods (Côté and Maljkovic 2010).

Local densities

Lionfish densities in the Atlantic are reported to be much higher than observed in their native range (Kulbicki et al. 2012). The first assessment of lionfish densities was provided by Whitfield et al. (2007), who reported an average of 21 invasive lionfish ha^{-1} across 17 locations off North Carolina in 2004. By 2008, the maximum lionfish densities observed off North Carolina were approximately 450 lionfish ha^{-1} , with mean densities around 150 lionfish ha^{-1} (Morris and Whitfield 2009). Green and Côté (2009) reported higher lionfish densities in the Bahamas in 2008, with mean lionfish densities of 393 ha^{-1} across three sites off New Providence. Similarly high densities have been reported by other studies in the Bahamas (Albins and Hixon 2011).

These densities are much higher than reported in the native range with Fishelson (1997) reporting ~80 lionfish ha^{-1} (assuming a mean search width of 10 m along the 1 km reef stretch), and also much higher than the density of *P. miles* located along the Kenyan coast (Darling et al. 2010). Grubich et al. (2009) reported less than 14 lionfish ha^{-1}, including all species of the *Pterois* and *Dendrochirus* on reefs of the Palau Archipelago. A regional assessment of *Pterois* spp. density from across the native Indo-Pacific region reports a maximum density of 26 individuals ha^{-1} (Kulbicki et al. 2012).

Reproduction

The pteroines, including *P. miles* and *P. volitans,* are gonochoristic (separate genders from birth) with males and females exhibiting minor sexual dimorphism during spawning (Fishelson 1975). The two genders are morphologically identical and thus cannot be distinguished visually. Males typically grow larger than females with the largest male lionfish recorded as 476 mm total length (J. Morris, Unpublished data).

Courtship has been described for the pygmy lionfish, *Dendrochirus brachypteru,* (Fishelson 1975) and zebra turkeyfish, *Dendrochirus zebra,* (Moyer and Zaiser 1981), a closely related genus to *Pterois*. Fishelson (1975) reported that courtship behaviors for *Pterois* spp. are similar to *Dendrochirus brachyopterus* in that the male and female circle each other, side-wind, follow, and lead one another beginning shortly before dark and extending well into nighttime hours. Following courtship, the female releases two buoyant egg masses that are fertilized by the male and ascend to the surface. Each egg mass contains approximately 10,000–20,000 eggs, depending on female size (Morris 2009). The eggs and later embryos are bound in adhesive mucus that disintegrates within a few days, after which the embryos and larvae become free floating (Morris et al. 2011).

Moyer and Zaiser (1981) reported that lionfish egg masses may be chemically defended based on observations of avoidance by some egg predators in aquaria. In contrast, a single sergeant major *Abudefduf saxatilis,* a common egg predator of the Atlantic, was observed feeding on a lionfish egg mass in the laboratory (J. Morris, Unpublished data). The seasonality of lionfish reproduction throughout their native range is unknown. Collections off North Carolina and in the Bahamas suggest that lionfish reproduce in all seasons of the year, approximately every 3–4 days (Morris 2009).

Early life history
Lionfish embryos hatch at the surface from a buoyant egg mass. Complete stage descriptions of *P. miles* and *P. volitans* larvae are not available. A few larval lionfish have been described, including five *P. volitans* larvae collected off northwestern Australia (Imamura and Yabe 1996) and one postflexion larvae collected off Cancun, Mexico (Vásquez-Yeomans et al. 2011). Developmental progression, typical of fishes that release pelagic eggs, includes larval dispersal at the surface or within the water column until settlement to the reef as juveniles.

Scorpaenid larvae exhibit two morphologically distinct groups characterized as "morph A" and "morph B" by Leis and Rennis (2000). Pteroine larvae are grouped among the "morph B" morphotypes, whose traits include: large head, relatively long and triangular snout, long and serrated head spines, robust pelvic spine, pigment confined to the pectoral fins (Leis and Rennis 2000), and postanal ventral and dorsal midlines (Washington et al. 1984). *Pterois* sp. meristic characters are reported as 12–13 dorsal spines, 9–12 dorsal rays, 3 anal spines, 5–8 anal rays, 12–18 pectoral rays, 1 pelvic spine, 5 pelvic rays, and 24 vertebrae (Imamura and Yabe 1996, Leis and Rennis 2000).

The size of *P. miles* or *P. volitans* larvae at hatching is unmeasured, but is likely to be approximately 1.5 mm as seen for *P. lunulata* (Mito and Uchida 1958, Mito 1963). The settlement age of lionfish in the Atlantic is estimated to be between 20–35 days with a mean of 26.2 days (Ahrenholz and Morris 2010).

Dispersal
Lionfish larvae and eggs are capable of dispersing great distances (Ahrenholz and Morris 2010, Vásquez-Yeomans et al. 2011). As a result, lionfish that recruit to any locale could be the result of spawning afar. Lionfish recruitment is expected to vary across marine habitats in response to variable physical oceanography and local environmental variation in habitat quality and resource availability. Long distance dispersal of lionfish occurs primarily during the pelagic larval phase, during which geostrophic and wind-driven currents transport the larvae. For example, lionfish eggs released in the Bahamas are capable of dispersing to New England via the Gulf Stream.

Larval connectivity models for reef fishes (e.g., Cowen et al. 2006) provide insight into lionfish larval dispersal and may be valuable for predicting the spread of lionfish, as evidenced by the recent establishment of lionfish in the Caribbean and Gulf of Mexico. Recent efforts by Vásquez-Yeomans et al. (2011) demonstrated the use of oceanographic models to hindcast dispersal pathways for lionfish larvae in the Caribbean. Although it is currently unknown whether lionfish recruit to specific habitat types or respond to environmental cues for settlement, recruits and adults have been observed in similar habitats in both shallow and deep water. Looking ahead, larval recruitment dynamics may be affected by regional stressors such as climate warming, which could result in a shorter larval duration stage and increased local recruitment (Côté and Green 2012).

Venomology
All lionfish spines, with the exception of caudal spines, contain apocrine-type venom glands. Each spine is encased in an integumentary sheath or skin and contains two

grooves of glandular epithelium that comprises the venom-producing tissue. Glandular tissue extends approximately three-quarters the distance from the base of the spine towards the tip (Halstead et al. 1955). The 13 dorsal spines, 3 anal spines, and 2 pelvic spines are capable of causing envenomation (Figure 2.4).

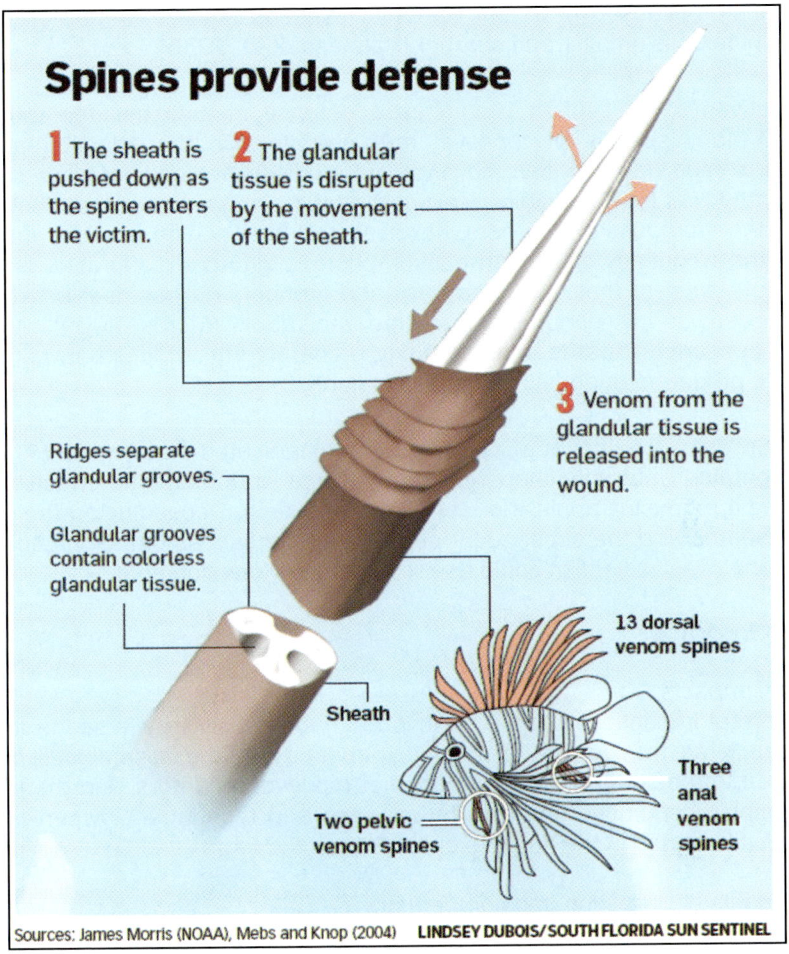

Figure 2.4 Graphic describing the lionfish sting morphology and mechanics.

Lionfish envenomation occurs when the spine's integumentary sheath is depressed as it enters the victim. This process tears the glandular tissue allowing the venom to diffuse into the puncture wound (Saunders and Taylor 1959). The toxin in lionfish venom contains acetylcholine and a neurotoxin that affects neuromuscular transmission (Cohen and Olek 1989). Lionfish venom has been found to cause cardiovascular, neuromuscular, and cytolytic effects ranging from mild reactions, including swelling, to extreme pain and paralysis in upper and lower extremities (Kizer et al. 1985, Badillo et al. 2012). Antivenom of the related stonefish (*Synanceia* spp.) is highly effective in neutralizing lionfish venom activity (Shiomi et al. 1989, Church and Hodgson 2002).

The severity of reaction to a sting depends on several factors: the amount of venom delivered, the immune system of the victim, and the location of the sting. Records of home aquarists stung by lionfish provide a comprehensive assessment of how lionfish stings affect humans (Kizer et al. 1985, Vetrano et al. 2002, Badillo et al. 2012). The probability of lionfish envenomation is higher when handling smaller-sized lionfish, because the venom glandular tissue is closer to the tip of the spine and the spine tip is smaller and sharper (Halstead et al. 1955).

The effectiveness of lionfish venom as a defense in their invaded range is in question. Many species of top-level reef predators in the Atlantic have been observed to feed upon lionfish, however, no studies have correlated predation on lionfish with lower local densities. Maljković et al. (2008) reported that lionfish were found in the stomachs of groupers, however, this observation provides no assessment of the frequency of lionfish consumption by grouper. Laboratory behavioral experiments suggest that black sea bass and groupers (Serranidae) actively avoid lionfish, even during periods of extreme starvation (Morris 2009). Additional research is needed towards understanding predatory interactions between lionfish and native predators during all life stages of lionfish.

Work by Sri Balasubashini et al. (2006a, 2006b) indicated that lionfish (*P. volitans*) venom contains antitumor, hepatoprotective, and antimetastatic effects in mice, suggesting a promising application for cancer research. Depending on the outcome of this research and the subsequent demand for lionfish venom, bioprospecting of venom from invasive lionfish could assist with fishery development.

Invasion Impacts

Ecological impacts

The ecological impacts of lionfish on invaded habitats will vary in space and time. Factors such as the density of lionfish, habitat type, the assemblage of native species within the community, abundance of top level predators, and the effects of oceanographic conditions on colonization rates and recruitment dynamics can all play a significant role in the impacts of lionfish on the reef.

Direct impacts of predation and competition

Albins and Hixon (2008) provided the first evidence for the impacts of predation by invasive lionfish, reporting a 79% reduction in fish recruitment on experimental patch reefs in the Bahamas during a five-week observation period in the presence of a single small lionfish. These experimental effects of predation appear to scale up to reef-level impacts. Sharp declines in prey biomass have been observed on natural Bahamian coral reefs, with prey biomass reduced by an average of 65% over a two-year-period (Green et al. 2012a).

Lionfish predation has been shown to affect fish species composition, with variable impacts realized across species and size groups (Albins 2012). The work of Lesser and Slattery (2011) also indicated ecological effects to fish communities on deep mesophotic coral reef in the Bahamas, which may have resulted in a phase shift to an algal dominated coral community at mesophotic depths, a result thought to be due to lionfish over-consumption of herbivores.

Lionfish occupy similar habitats and consume similar prey to many species of native fish predators (Morris and Akins 2009, Green et al. 2012a) and macro-invertebrates. Competition with lionfish may affect the behavior, distribution, growth, survival, and, ultimately, population size of these ecologically similar native species. An experiment on coral reefs found that lionfish grow significantly more quickly and consume prey at rate far faster than the native predator Coney grouper (*Cephalopholis fulva*), which raises concerns that lionfish could outcompete some native predators for food resources on invaded habitats (Albins 2012).

Indirect effects on marine community structure and resilience

The overall ecological impacts of lionfish can be framed in the context of impacts to biodiversity and community composition and function, and thus resilience of reef systems. Whether Atlantic systems are biotically resilient, or even resistant, to the effects of lionfish is not known, but this will vary by location in relation to the composition of the biotic community and physical oceanographic features of each location. Albins and Hixon (2011) provided a summary of worst-case scenarios for the lionfish invasion, which highlighted the many life history and ecological traits that make lionfish highly invasive (Morris and Whitfield 2009). The direct effects of lionfish on native fish and invertebrate communities (i.e., through predation and competition) may indirectly affect the base of marine food webs, if the species consumed or outcompeted by lionfish influence the diversity and biomass of benthic communities. In actuality, the direct and indirect impacts are likely to be exacerbated by the degraded condition of Atlantic reefs, owing largely to overfishing, pollution, and climate change. For example, the effect of predation by lionfish on their prey base is predicted to worsen with future climate warming in the region (Côté and Green 2012).

Socioeconomic impacts

The socioeconomic impacts of lionfish remain largely unquantified, but have the potential to be severe. Vulnerable sectors include fishing and tourism economies, which are critically important to many Caribbean and Atlantic nations. For example, lionfish predation on economically important species, such as juvenile serranids (Morris and Akins 2009), could cause a decrease in landings, hamper stock rebuilding efforts, and slow conservation-based initiatives.

Further, lionfish impacts on tourist recreational activities have been observed. Some locations have posted warning signs (Figure 2.5) advising of the potential for lionfish envenomation. As lionfish densities increase, so too does the risk of envenomations. It is unknown whether increasing lionfish densities will reduce recreational activities and cause economic hardship. This is likely to be dependent on factors such as the local density of lionfish, the rate of human encounters with lionfish, and the effectiveness of education and outreach. Research that monitors lionfish socioeconomic impacts on the fishing and tourism economies is needed.

Figure 2.5. Lionfish warning sign on a beach in the Bahamas.

REMAINING QUESTIONS

Research priorities for lionfish are likely to differ with the specific needs of each locale and the stewardship responsibilities of each organization. Further, the need for research coordination on lionfish is crucial. Sharing of new ideas, approaches, successes, and failures is critical to development of regional standards for lionfish control.

A number of research priorities were identified during the 2010 Cancun Lionfish Workshop. These included:

i) Quantify the socioeconomic impacts of the lionfish invasion,
ii) Model projected expansion of the invasion, including larval dispersal,
iii) Conduct applied research on lionfish bycatch,
iv) Conduct applied research to target lionfish effectively in deep water,
v) Standardize research methods,
vi) Determine the effectiveness of removals,
vii) Highlight the risk, or lack thereof, from ciguatera associated with lionfish,
viii) Explore lionfish connectivity,
ix) Conduct spatial monitoring of lionfish,
x) Investigate sterilization as a potential control method, and
xi) Investigate invasive and native species interactions.

In May 2011, a focus group was organized to identify knowledge gaps and coordinate research to inform the management of invasive lionfish in the western Atlantic at the 2nd International Marine Conservation Congress (IMCC), in Victoria, British Columbia. This focus group developed additional priorities with finer detail, including the following:

i) Biotic resistance to the invasion:
   - Is there evidence that some marine communities in the Atlantic are resistant to invasion by lionfish?
   - If so, what species interactions promote resistance and what is the frequency of these interactions across the region?
   - Are ecologically "intact" communities more resistant to the lionfish invasion?
   - Can native predators be trained to eat live lionfish and would this predation be ecologically significant?

ii) Do biotic interactions limit lionfish populations in their native range?

iii) Which stage of lionfish life history most limits population size?

iv) Which species, if any, consume larval and post larval lionfishes in their native range? If so, do similar species inhabit the Atlantic and what is the potential for them to consume lionfish?

v) How does the fitness and survival of invasive and native lionfish populations compare, and is there a genetic basis for differences between the two?

vi) What habitats are lionfish larvae recruiting to? Are there specific environmental and biotic cues for settlement?

vii) How quickly do lionfish populations increase after first colonizing an area, and what factors affect the rate?

viii) What is the density and distribution of lionfish in mangrove, seagrass, estuary, hardbottom, and deep-sea habitats?

ix) What is the ecological impact of lionfish on native marine communities in mangroves, seagrass, estuary, hardbottom, and deep-sea habitats in the Atlantic?

x) What is the effect of lionfish predation in fish nursery areas on adult fish populations?

xi) Will lionfish population increases ever level off? Is there density dependence?

xii) The costs and benefits of fishing lionfish in the Atlantic:
   - At what scale is lionfish fishing currently occurring?
   - Is fishing lionfish a viable option for population control?
   - What are the socioeconomic effects of promoting lionfish fisheries?

xiii) What is the effect of lionfish predation on benthic communities?

xiv) Where do lionfish spawn? What environmental factors influence reproductive activity?

xv) Are there source and sink populations of lionfish, and how can we use this information to direct local control efforts?

RESEARCH RESOURCES

Resources are available for researchers looking to answer questions relating to the lionfish invasion. Well over 30 peer-reviewed journal articles have been published on the lionfish invasion, covering biology, ecology, control, and distribution. As with most research topics, researchers new to lionfish should be wary of Internet sources and other non-peer-reviewed media (including listservs), as these sources contain much misinformation. Careful researchers should make it a practice to always check sources of information.

Lionfish research symposia have been held yearly since 2008 during the annual conference of the Gulf and Caribbean Fisheries Institute. Proceedings from these symposia can be found at: http://www.gcfi.org/Lionfish/Lionfish.html.

During an International Marine Conservation Congress Focus Group Session in 2011, a list of researchers and topics was compiled to promote coordination among researchers. Information about this session can be found by contacting Stephanie Green at: stephanie.j.green@gmail.com.

A Lionfish Web Portal (see http://www.GCFI.org) is presently being developed and will serve as a clearinghouse for information on lionfish in the Caribbean. Additional online resources are provided below:

NOAA Lionfish Website:
http://www.ccfhr.noaa.gov/stressors/lionfish.aspx

USGS Lionfish Factsheet:
http://nas.er.usgs.gov/queries/factsheet.aspx?speciesid=963

REEF Lionfish Research Program:
http://www.reef.org/programs/exotic/lionfish

Oregon State University, Hixon Laboratory:
http://hixon.science.oregonstate.edu/content/highlight-lionfish-invasion

CHAPTER 3

EDUCATION AND OUTREACH: *Building Support and Expertise*

J. Lad Akins

Successful Education and Outreach (E&O) programs can help shape public perception, enhance constituent involvement, and secure government support and funding. Because effective lionfish control programs encompass all of these components, they will benefit greatly from well thought-out and designed E&O activities. In general, the earlier appropriate E&O activities are implemented, the more broadly they are supported.

This chapter provides information on E&O strategies aimed at increasing support for programs addressing the lionfish invasion. Topics include developing E&O priorities, key messages, program considerations, communication strategies, and examples of messaging and outlets.

THE EARLIER, THE BETTER

By forming and implementing an E&O program *before* a lionfish invasion, managers can greatly aid their efforts to manage the problem. In two successful examples — the Bonaire Marine Park (Stichting Nationale Parken Bonaire/STINAPA) and the Florida Keys National Marine Sanctuary (FKNMS) — education and outreach activities were initiated prior to lionfish colonizing their waters. These management organizations targeted their proactive outreach activities to the general public, dive industry, health and medical industries, educational institutions, and the media. They developed outreach materials, including posters and stickers, and control plans; and they held workshops in advance of the invasion (Figure 3.1).

The STINAPA and FKNMS activities ensured that all user groups in each respective location were equipped with accurate information and were aware of protocols for responding to lionfish sightings. Because of these advance efforts, community awareness in both locations was relatively high and misconceptions were addressed prior to the invasion. As lionfish began to colonize local areas, the respective communities supported removal activities. They continue to be among the most supportive communities in the region.

Figure 3.1 Sticker used by the Florida Keys National Marine Sanctuary to encourage reporting of sightings.

E&O TO SUPPORT CONTROL PLANS

The primary mission of most lionfish control plans is to minimize the impacts of the invasion. Impacts can take many forms, including disruptions to marine communities (environmental), fishing and tourism activities (economic), and the safety of those encountering or collecting lionfish (human health).

The level of E&O effort depends on the specific goals and objectives of the program. For instance, managers who dedicate time to developing solid mission statements, goals, objectives, and strategies will strengthen their E&O efforts and help facilitate a stronger overall program. What's more, those who involve stakeholders from broad sectors (e.g., media, education, research, management, political office) will also help their programs by identifying key audiences and strategic methods for reaching them.

Some examples of outreach goals for minimizing lionfish impacts are:

i) Increase removals of lionfish to reduce local populations,
ii) Increase public awareness of lionfish impacts to generate support for effective management,
iii) Maximize efficiency in utilizing human and fiscal resources,
iv) Increase political support for lionfish management and control,
v) Minimize health risks to the general public and those handling or collecting lionfish,
vi) Link lionfish research to management strategies, and
vii) Prevent future introductions of additional lionfish or other non-native species.

Who Owns the Lionfish Problem?

Everyone! The impacts of the lionfish invasion are likely to be broad, spreading across many intricately linked sectors. The effects are also likely to be long term in nature and will require a deep understanding of how working together can provide the strongest benefit.

No one person or office has the capacity to deal effectively with the broad impacts of this invasion. Like a fire burning at the edge of town, the invasion requires that everyone chips in to address the threat, as they are able.

DEVELOPING COMMUNICATION STRATEGIES

In developing communication strategies, managers might find it helpful to first identify what they want the E&O activity to accomplish. Ask the question: "What would successful communication look like or result in?" If the answer is "increasing the number of people who realize lionfish are a threat to native marine systems or economies" or "increased reports to the reporting hotline" or "more people eating lionfish," the manager should select target audience(s) and messages to achieve that desired outcome. Once the target audiences have been identified — e.g., fishers, divers, politicians, the general public, and/or businesses — then messages can be developed that are specific to that target group. It is important to select broad overarching messages aimed at the groups. See Figure 3.2 for an example of how to determine target audiences and Appendix 1 for a list of some lionfish-specific strategies currently in use.

Developing an E&O Program

The following steps may be useful for creating an E&O program for local or regional lionfish control efforts:

i) Identify your E&O goals; e.g., to increase the number of divers proactively removing lionfish.
ii) Prioritize your E&O goals according to your overall program objectives.
iii) Identify your audience(s) for each goal.
iv) Craft goal-specific messages appropriately to each audience.
v) Identify communication tools and audience specific messages; e.g., social media outlets could work well for certain audiences, but might not be effective at reaching others without Internet access.

Importance of clear and credible messaging

When developing messaging for outreach efforts, managers should remember that not all outlets are equal. Messaging should be tailored to suit both the media outlet and target audience. Print, radio, television, documentaries, and public and political forums are all key messaging venues or "tools," but each one has its differences and limitations in format, content, style, and length. Different versions of the same general message can be used for different groups; however, it remains important to remember that the audience and the desired behavioral change will dictate which of the many messages are relevant.

It is also important that managers and programs stay "on message," especially when working with the media or in public forums. It is easy to get sidetracked by a question or comment, but it is important to come back to the main message.

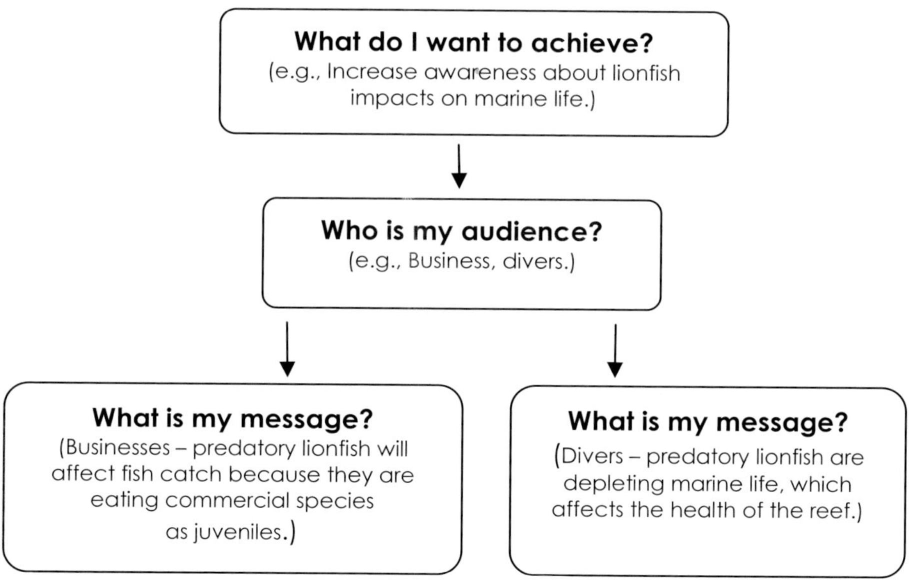

Figure 3.2 Example of a decision-tree approach for developing a communication strategy.

Key messages
Most people have fairly short attention spans. While managers may want to convey every known piece of information about lionfish and the invasion, it is important to distill and refine the material into a subset of key messages. These key messages should be designed to elicit the changes in perception and behavior necessary to support program goals. It is important that all partners are consistent with messages delivered in their E&O activities.

It is also important to remember that key messages will vary by audience and may change over time in response to various control strategies. Effective programs will periodically re-evaluate and update these messages.

Vetting information
The information to be disseminated through E&O activities must be accurate. In this age of mass communication and public media, messages are sometimes disseminated before their accuracy is verified. Misinformation or exaggeration through hyperbole can damage the credibility of E&O programs and hinder the success of management. It often takes only one small piece of incorrect information for an entire message or outreach campaign to lose credibility. Managers should take the utmost care to properly verify all information used or distributed in outreach efforts.

Sources of credible information
Accurate information on lionfish and the lionfish invasion is widely available. Examples of credible information sources include peer-reviewed scientific publications, direct quotes, and the websites of peer-reviewed researchers. When in doubt, managers should be sure to verify the messages with a second opinion or outside source — prior to distribution. (For more information on credible sources of information, see Chapter 7.)

Outreach Outlets

Standard media venues — such as print, radio, television, film documentaries, and public and political forums — are all key messaging outlets or "tools." Managers can also think beyond these standard formats to come up with ideas that will work specifically for their management program, their community, and their target audience(s).

Figure 3.3 Pamphlet produced and distributed in Cuba to educate primary school children about lionfish.

In an example of outreach to a younger audience, the Acuario Nacional de Cuba has developed a series of educational pamphlets entitled Conozcamos El Mar, focusing on marine organisms and related issues and targeting school-age audiences (Figure 3.3). The pamphlets run about 25 pages and include relevant information presented in audience-appropriate methods — cartoons, puzzles, characters, etc. They are distributed to public schools throughout the country. Volume 12 of the series focuses specifically on lionfish.

By prioritizing media outlets according to the key audiences they reach and the return on their investment — namely, the time and money spent — managers can maximize efforts. Creative venues can often reach traditionally underserved audiences; however, the investment should be weighed in relation to the benefit of reaching those audiences. Tables 3.1 and 3.2 describe the costs and benefits of various media outlets.

Key Lionfish Messages

Lionfish are an invasive species and are detrimental to native systems.
Lionfish are not native to western Atlantic waters and are capable of causing negative impacts to native marine life, ecosystems, economies, and human health.

Aquarium releases as source of the invasion.
Genetic research and monitoring of lionfish distribution suggest that the source of introduction is likely to have been multiple releases of aquarium specimens off the coast of southeast Florida.

Eggs and larvae are transported via ocean currents.
Lionfish are distributed to new areas via dispersal of their eggs and larvae by ocean currents.

Invasion progresses rapidly.
Even though initial sightings of lionfish in a new area can be sporadic over time and space, the invasion progresses rapidly. Most countries have experienced invasion progression from the first sighting to multiple lionfish occurring on most sites in less than two to three years.

Impacts may be severe.
Recent research indicates that lionfish impacts can be severe and cross broad spectrums of the environment, including economically important species like juvenile grouper and snapper and ecologically important species like grazers and cleaners (e.g., parrotfish, cleaner shrimp and fish).

Natural predation is not controlling.
While some incidental predation or conditioned feeding on captured lionfish may occur, it appears that there are no controlling predators of lionfish in this region.

Community involvement is necessary.
To effectively address the lionfish invasion, we must develop wide-scale support and the involvement of the local communities.

We can make a difference: Local control can be effective.
Local control efforts, including adopt-a-reef type programs and the development of food-fish markets are showing success. Areas that promote and conduct regular removals are showing fewer lionfish than non-removal areas, though removals will need to be long-term in nature due to recruitment of lionfish from upstream populations (see eggs and larvae message above).

Lionfish are edible.
In their native range, lionfish are considered a delicacy and are consumed regularly. Close relatives of the lionfish, such as rockfish (Sebastes spp.) and the blackbelly rosefish (Helicolenus dactylopterus) — which are also scorpaenid fishes — are harvested commercially. Human health concerns associated with eating fish, such as ciguatera and mercury levels, have not been determined for all locations, though some lionfish have been found to carry the ciguatera toxin in known hotspot locations. Managers should implement the same caution for lionfish as they do for other reef fish.

Venom does NOT equal poison.
Lionfish possess venomous dorsal, anal, and pelvic spines for defense. The meat of lionfish does not contain poison.

Eradication is not likely.
Under current technologies and considering the spatial extent and severity of the invasion, eradication is not a likely outcome. Honest dialogue regarding this issue is important in developing accurate and achievable outcomes and subsequent strategies. Additionally, statements regarding the possible eradication or prevention of lionfish establishment will set up a situation of distrust and damage credibility as these goals are unmet.

Table 3.1 Examples of outreach outlets are provided below with samples of targeted audiences, message type, advantages/disadvantages, and development times for each.

Outlet/tool	Audience	Message types	Advantages	Disadvantages	Development time
TV, radio	General public	New information Events	Free Large audience Can be visual	Live interviews offer little chance for review Low control over message Short segments Little repeat viewership	Short
Documentaries	Select audience according to interest	Detailed information	Managers can often help direct message or story	Significant delays between filming and broadcast	Very long
Magazines, newspapers	General public	New information Events	Free Audience size (varies) Opportunity for review Multiple readership	Text-based format requires focused attention of reader	Short
Printed outreach (flyers, stickers, pamphlets)	Targeted audience (depending on distribution)	Ongoing messages	Long-lived Select distribution can target specific audience or locations Can contain detailed information	Limited audience Can be defaced Difficult to modify with updated information	Long
Websites	Select audience according to interest	Ongoing	Can provide wealth of information (links)	High maintenance May require specific expertise	Very long (to develop initially)
E-mail	Select subscribers	Updates Events	Short message length Text based	Can be easily discarded without reading	Short
Public forums	Select audience according to interest or venue	Detailed information	Opportunity to engage public in Q&A	Misinterpretation of information Opens forum to special interests	Moderate
Social media	Select audience according to subscription	Very brief info bytes	Rapid and frequent Can provide links to detailed info Easily circulated	May get lost in high volume media Low longevity	Very short
Listservs	Select audience according to subscription	Detailed professional briefings	Can provide links to detailed info Easily circulated Often archived for later searches	May get lost in high volume	Short

Table 3.2 Sample outreach outlets matched with media type, number of people reached, and relevance of the audience. Readers may consider using this table to prioritize messaging by matching outlet with importance of audience and number of persons reached.

	Outlet	Media type commonly utilized	Number of persons reached	Relative importance of audience reached
Private				
	Resorts	Printed materials	High	Low
		Forums	Low	Low
	Tour/dive operators	Printed materials	Med	High
		Forums	Med	High
	Commercial fishers	Printed materials	Low	High
		Forums	Med	High
	Recreational fishers	Printed materials	Med	Low
		Forums	Med	Low
	Restaurants	Printed materials	Med	High
	Aquaria	Printed materials, forums, videos	Low	Med
	Cruise lines	Printed materials, video	High	Low
	Pet trade	Printed materials	Low	Med
	Physicians/medical personnel	Printed material,s forums	Low	High
	Citizens	Printed materials, media, forums	High	Med
	Chambers of Commerce	Printed materials	Low	High
	Funders	Forums	Low	High
Government				
	Political leaders	Printed materials, forums	Low	High
	Enforcement	Printed materials, forums	Low	Med
	Resource managers	Printed materials, forums, listservs, web	Low	High
	Health	Printed materials, forums	Low	Med
	Tourism	Printed materials	Low	Med
	Customs	Printed materials	Low	Low
Social/Civil/NGO				
	Educators and students	Printed materials, forums	High	High
	University researchers	Forums	Low	High
	Churches	Forums	Med	Med
	Conservation NGOs	Printed materials, forums	Med	High
	Clubs	Forums	Low	Med
	Philanthropic foundations	Printed materials, forums	Low	High

CHALLENGES

Accurate information and a thorough plan are the foundations of a solid E&O program, but even a good outreach plan can be de-railed by a few minor issues. Some easily avoidable pitfalls are described below.

Lack of coordination among points of contact
Designate one or two people for each media release to be the primary contacts. If possible, utilize the same contacts for all messaging. These contacts typically field general questions and direct media to the appropriate person for more information.

Unsustainable (static) messaging
Avoid sending the same (or similar) media releases repeatedly to the same outlets. Be creative and come up with new angles or elements.

Unbalanced messaging
Avoid unbalanced messaging between human health and ecological impacts. For example, human health risk may differ in priority compared to ecological impacts; therefore, it is important to note these differences.

Hurricane Andrew Aquarium Release Story

In 1995, a brief in an American Fisheries Society newsletter, based on third-hand information, reported that a bayfront aquarium had been destroyed during Hurricane Andrew three years earlier, releasing several lionfish into Biscayne Bay. At the time, this brief generated little attention. As the lionfish invasion progressed, however, media began to circulate this story, citing the single-storm event as the source of the invasion. The author has since commented that the information was never intended to document the source of the invasion and has never been verified. Recent genetic work has also disproven this single-release theory and points to multiple release events that have included at least 10 lionfish (Freshwater et al. 2009, Betancur - R. et al. 2011).

Read more about the lionfish-hurricane story in Science Insider at:

http://news.sciencemag.org/scienceinsider/2010/04/mystery-of-the-lionfish-don't-bla.html?ref=hp

Similarly, unbalanced messaging regarding human consumption can quickly de-rail local control efforts. For example, while the risks of ciguatera from consuming lionfish may be worth noting, there is little information on the relative risk compared to that posed by consuming native species. Therefore, the message should be balanced, and provide similar caution for ciguatera in lionfish as is provided for ciguatera in native reef fish.

Inaccurate information
Be sure to validate and verify information before passing it along. Common inaccuracies, such as the source of the lionfish introduction or location of venomous spines can confuse audiences and can place people at risk.

Inappropriate visual aids
Use images that support the message. For example, when describing lionfish impacts managers should use images depicting the significance of the invasion, such as a picture showing high densities of lionfish or degraded fish communities. A beautiful image of lionfish in its native range could create an affinity for lionfish

among some viewers. Also, as there are many different species of lionfish in the native range, the use of incorrect species when describing the invasion may cause confusion.

MEASURING SUCCESS

Measuring changes in public perception and the effectiveness of messaging is difficult. Feedback is important in determining the direction for increasing success. There are a few simple tools that managers can use to determine how well their outreach programs are working. They include:

i) *Tracking* — Responses to e-mails, calls, or personal feedback can be tracked and summarized to determine where the respondent heard about the issue. A simple notepad next to the phone will often suffice.

ii) *Participation* — Keep track of how many people attend specific events or public forums, and relate that back to the outlet and messaging used.

iii) *Short questionnaires* — Questionnaires and/or surveys can be useful tools, though special training or skills may be required to provide valid results. Administering questionnaires through face-to-face contact or via e-mail can provide valuable information on how the perceptions and behaviors of target audiences are changing in response to E&O efforts. Be cautious about survey length and in the use of questions that may lead the respondent towards a specific answer.

CHAPTER 4

CONTROL STRATEGIES: *Tools and Techniques for Local Control*

J. Lad Akins

The source of the lionfish invasion will never be completely understood, but the fact that lionfish are threatening a vast portion of the tropical and subtropical western Atlantic is clearly evident. The impacts of lionfish are likely to resonate through economies, human health, and tourism, and they will include both direct and indirect ecological components over the long term. The goal of any control program will focus on minimizing impacts.

Whether through developing new technologies, buying time for natural control to evolve, or reducing the severity of impacts, virtually all of the strategies described in this chapter focus on reducing the size of lionfish populations and their corresponding impacts. It is going to be up to all stakeholders to work together to address the invasion. Activities supporting that goal are numerous, but the approach is simple: increase awareness, enact effective removals, and maximize efficiency.

This chapter identifies and discusses many different strategies with examples of how to effectively accomplish these objectives, including a framework for developing a control plan, tools, techniques, spatial scales and frequencies for control, and methods for maximizing efficiency.

LOCAL REMOVALS CAN BE EFFECTIVE

While region-wide eradication and prevention are unlikely scenarios, recent research and anecdotal information indicate local control efforts can be highly successful in managing lionfish densities and minimizing impacts. Development and support of removal efforts not only reduce effects on native populations, they also buy time for development of new technologies and may allow for natural control mechanisms to evolve. It is important to remember that for every lionfish removed from the water, a reduction in predation on native species is realized. In already stressed marine systems, even a small level of decreased impact could result in significant long-term benefits.

Local removal efforts can reduce lionfish numbers, but resources generally limit the geographical scope of a management area. That's why it is important to identify and prioritize sites for removal. Key areas such as marine protected areas (MPAs), high visitation or tourism areas, spawning aggregation sites, vulnerable nursery sites, and/or other areas may be deemed high priority for removal efforts. Removal plans should consider these priority areas in relation to available resources.

How Many Lionfish Should Be Removed?

Recent publications by Morris et al. 2011 and Barbour et al. 2011 indicate that large numbers of lionfish (27% of the adult population per month in the Morris model and between 15–65% each year in the Barbour model) would need to be removed regularly to elicit a decrease in the overall population. Large-scale removal efforts with this goal are simply not feasible with current technology, but recent efforts in the Bahamas, Mexico, Cayman Islands, and the Florida Keys indicate that locally directed control at specific sites can be successful in keeping lionfish densities low in removal locations.

Taking Advantage of Resources

While fisheries or resource managers are likely to be those charged with organizing and directing lionfish control programs, many resources outside of the government are available. Working in partnership with resource users, stakeholders, non-governmental organizations, and neighboring countries, managers can add significant capacity to control efforts. Many partners are eager to assist and can enhance and broaden government-led programs. By working together, programs not only increase their removal efforts, but they also form valuable partnerships that may benefit other programs and issues. These partnerships — including those that involve volunteers — are important for removal success. Effective partnerships, however, do not run by themselves. They require dedicated coordination and communication.

An Overview of Control Plans

Advantages of having a plan

Many countries in the lionfish-invaded region have already developed plans for lionfish control. The strategies, tools, and techniques used by each region depend on factors such as local environmental conditions, regulatory and societal structure, and human and financial resources. Common themes run among most plans, yet consideration of new, alternative, and/or non-traditional activities may be beneficial — even in already developed control plans.

It is easy to recognize the challenges facing marine resource managers and government officials. For instance, addressing the new threat of invasive lionfish has not been in the long-term strategic plans of most locations, and it places new burdens on thinly stretched resources. Nonetheless, developing plans for how to address the persistent lionfish issue is as important as plans for addressing other threats to economies and environments. Many of the same elements employed in other conservation issues can be used in developing lionfish control plans, and effective lionfish control strategies often dovetail with existing management and conservation activities.

Some advantages of developing a lionfish control plan:

- i) Provides direction for pro-active (as opposed to re-active) management,
- ii) Facilitates partnership building,
- iii) Provides provisions for public involvement and enhances public perception of management efforts,

iv) Outlines effective allocation of resources,
v) Enhances understanding of invasion impacts by defining research priorities, and
vi) Increases coordination and communication among stakeholders (reduction in stakeholder conflicts).

Considerations for plan development
Good control plans should clearly define roles and responsibilities, provide regulatory mechanisms for implementation or modification of new and existing laws and policies, and offer adaptive management frameworks. See Chapter 6 for additional framework, roles and responsibilities, and regulatory descriptions.

Designating leadership and responsibilities
Leadership designation is a key component of successful control plans. A strong lionfish control plan will outline who is in charge and where other groups and stakeholders fit into the organizational structure. Organizational trees (Figure 4.1) can be effective at identifying key roles and responsibilities. Resource managers will commonly experience a divergence of opinions on how to approach the lionfish invasion, and having a well-understood and transparent chain of command will facilitate coordination and inclusion of key stakeholders and user groups.

Developing priorities
Another component of control plans is to develop clearly defined priorities. Priorities may be best accomplished by including a diverse group of stakeholders, with differing areas of expertise, to capture broad-scale community interests. Identifying and understanding available resources — including human, fiscal, and logistical — is critical in developing priorities. Efforts to better understand resource vulnerability to lionfish impacts and to identify priority locations for lionfish control can then be used to match available resources and capacity.

Figure 4.1 Sample organizational tree describing one possible organizational structure of a lionfish control program.

Developing control strategies
When developing control strategies, resource managers will discover a few considerations that will go a long way in helping to implement realistic and effective efforts. Some of these include:

i) Working within existing capabilities,
ii) Tailoring strategies for local needs,
iii) Consideration of outside influences (recruitment variables such as water circulation, seasonality of spawning, year-class strength variability, nearby densities, benefits from neighboring control efforts, invasion equilibrium dynamics), and
iv) Working within existing regulatory frameworks.
    - Defining lionfish within existing regulations
    - Modification of existing regulations
    - Enacting new regulatory measures

Measuring success
In any control plan, an honest evaluation of successes and challenges is important. Building in review processes, setting realistic goals and expectations, and evaluating how well objectives were met are often overlooked — but critically important — components of successful plans. Resources managers should identify early in their planning how they will measure success. Some benchmarks may include: the number of lionfish removed, the number of lionfish remaining, the number of areas cleaned of lionfish, and/or the condition of the reef system. For more information on measuring success through monitoring, see Chapter 5.

Control Plan Components

Reporting and documentation
Reporting is an important component of any control plan as it helps to direct removal efforts and gauge the effectiveness of control programs. Reporting and documentation strategies can vary widely across regions and organizations; however, a single entity or contact is best to prevent confusion and duplication of effort. Obtaining reports during the initial phase of an invasion can occur with little relative effort, as the general public is typically excited and willing to assist. As the invasion progresses, however, the public can become desensitized and report less often. This may occur as the result of overwhelming lionfish densities or the public's rationalization and acceptance of the problem. Thus, outreach on the importance of reporting may become increasingly necessary during the later stages of the invasion. Without accurate reporting, resource managers can find it difficult to determine the effectiveness of control. Conversely, during advanced stages of the invasion, reporting can aid in focusing removal efforts on high priority areas.

Standardizing reporting
Standardized reporting mechanisms can be most effective for encouraging continued and accurate reporting. Phone, e-mail, and Internet should all be considered and consensus reached among agencies and partners on reporting protocols. The reporting format can be designed to follow regional protocols, and local reports can then be forwarded to regional nodes for display and dissemination. Additionally, user-based reporting tools can be developed to encourage and direct removal efforts on a real-time basis.

USGS maintains an international database of non-indigenous aquatic species that includes factsheets, distribution maps, and early notification of sightings. This database has been the primary repository of lionfish sightings and continues to be one of the most widely used tools in depicting the timing and scale of the invasion. Lionfish sightings should be reported to http://nas.er.usgs.gov/. For additional reporting forms see http://www.reef.org/programs/exotic/report — or http://nas.er.usgs.gov/sightingreport.aspx.

Follow-up actions
During the initial stages of the invasion, early detection/rapid response (ED/RR) is widely used to refer to the process of reporting and removal. It is important not only to encourage removals, but also to provide feedback on the response effort, to acknowledge the responders and provide further incentives, and to highlight the successes of these removals. These efforts will help build strong partnerships and pathways for future removal efforts.

Reporting considerations

i) Align reporting criteria with research needs and priorities.
ii) Maintain national and regional nodes for reporting, including a centralized database and single sources for information.
iii) Develop appropriate reporting and data management frameworks to incorporate different reporting methods and levels of informational detail.
iv) Simplify the reporting approach/method for volunteers and others.
v) Link collection to reporting for research needs.
vi) Link reporting to education (locally, nationally, and regionally).

COLLECTION AND HANDLING

The collecting and handling section of a control plan should be explicit in outlining who, what, when, where, and how removals are to take place. Through detailed and thoughtful design, programs can avoid ambiguity in allowable activities, policies, and procedures, and can facilitate effective removal efforts. There are many different methods of collecting and handling lionfish (Appendix 2), as well as myriad logistical considerations. The information below provides an overview of the most commonly used tools and techniques.

Defining personnel
A first consideration in designing control programs includes assessing available personnel. There is often strong interest, across many sectors, in aiding with lionfish removals. Identifying appropriate removal personnel and protocols for each sector will aid in a more organized and effective removal effort.

Government agencies
Government agencies often have a role in responding to environmental issues. Assigning a point person in the appropriate agency to represent the government and manage the overall response program is important. In addition to any lionfish-specific programs that are developed, there may be other agency activities that could easily incorporate a lionfish component without significant additional resources. Identifying who is doing what and defining what activities can be undertaken by government staff can be an effective way of providing leadership in addressing the invasion.

Professionals
Many projects are undertaken in coastal marine waters that involve underwater professionals and researchers. The professionals undertaking these projects may be an excellent source for both reporting and assisting with lionfish removals, during their normal activities. It is important to remember, however, that even minor removal efforts can expend resources and shift effort away from the core focus of their work. Solid communication with researchers should include discussions on what, if any, contributions they may be able to make to control efforts.

Partners — NGOs
Many non-governmental organizations (NGOs) are organized specifically to aid in issues affecting social or environmental sectors. Inviting and encouraging their participation in addressing the lionfish invasion may prove to be one of the most beneficial plan components in terms of human and fiscal resources. Care should be taken to provide solid direction, however, and to clearly define roles and responsibilities. Most NGO partners will also include large numbers of volunteers; therefore, defining a clear mission and role for the NGO is critical to partnership success.

Stakeholders — divers, fishers
In many areas, the group of stakeholders most affected by the invasion includes fishers and divers. Often, they are the first to encounter lionfish and the most capable of providing removal support. Key elements for engaging these stakeholders include education and outreach, training, and the provision of incentives for removal efforts. It is important to remember that not all incentives are monetary. Many stakeholders can be motivated to aid in removal efforts after learning the negative effect lionfish may have on their livelihoods and the integrity of their reefs.

Volunteers
While many volunteers will come forward under the leadership of NGO partners, there may also be significant volunteer interest by unaffiliated individuals. These may include visitors, tourists, concerned members of the public, individual members of government, and fishers and/or divers (acting outside of their normal work time). Managing and directing this interest can be a time-consuming and personnel-heavy undertaking; however, these individuals may also be some of the best supporters of lionfish control plans. If unaffiliated volunteers are going to be aiding in control efforts, proper training and supervision may be required and caution exercised around sensitive habitats. Further, responsible parties should consider liability concerns associated with handling venomous fish.

Collection Strategies

There are many methods currently used for removing lionfish from invaded waters. Based on wide-scale input from around the region and expert opinion from those who have considerable experience in removals, some methods seem to be more prevalent and effective than others. Managers should consider all methods and select the most effective and appropriate approach for their location. The information below provides a description of various methods.

Natural control
One of the most common questions about the lionfish invasion is what controls lionfish population growth in their native range and what may control them in the

Atlantic. Although it is clear that lionfish are rare-to-uncommon throughout their native range (Kulbicki et al. 2012), at this time, the mechanisms that control lionfish in their native range are not understood. Much attention has been given to the potential predation on juvenile and early adult lionfish by grouper and other top-level predators as a mechanism for population control in the Atlantic. It is unclear if predation on juvenile or adult lionfish may have a significant effect on controlling their population.

A recent report by Mumby et al. (2011) identified lower lionfish densities in the Exuma Land and Sea Park, Bahamas, compared to areas outside the park. Grouper densities were higher inside the park and were hypothesized to have contributed to the reduced abundance. There had been substantial lionfish culling efforts in the park prior to this study and no direct observations of the rate of predation by grouper were provided. Observations and experiments by other researchers have found that grouper seldom attack lionfish. More information is needed to better support this correlation.

For most species, mortality is highest during the egg and larval stages and diminishes with increasing age and size. It is unknown whether incidental predation on lionfish will increase over time, but it is anticipated that any predation pressure significant enough to control lionfish densities will be slow to emerge.

Other natural controlling factors could include competition, parasites, diseases, and/or genetic disorders. To date, parasites on lionfish in the invaded range have been found only rarely as compared to levels of parasites on native range lionfish or on other native species in the invaded range. Genetic deformities have been observed in lionfish in the invaded range, but surprisingly, these fish have been adults and were able to survive even with fin and/or other potentially inhibiting deformities. Disease is a potential controlling factor that is being examined at the time of this publication.

Teaching Top Predators to Control Lionfish?

Divers in some parts of the Atlantic have attempted to entice top predators (sharks, barracudas, grouper, snapper, and eels) to consume captured lionfish in the hopes that predators will learn to hunt and prey upon lionfish naturally. To date, there is no conclusive evidence that native predators are learning to prey upon lionfish through this practice. Unexpected effects of fish-feeding activities include aggressive changes in predator behavior during encounters with divers. Recent incidents include severe injury to a lionfish collector's hand from a barracuda bite (Figure 4.2) and harassment of other collectors by sharks and large groupers. These developments have proven the practice of feeding lionfish to predators as dangerous and counter-productive. In addition to serious human injury, the practice could also lead to injuries of top predators through diver responses to attack and/or increase in the ease of capture of predators that are conditioned to associate humans with food.

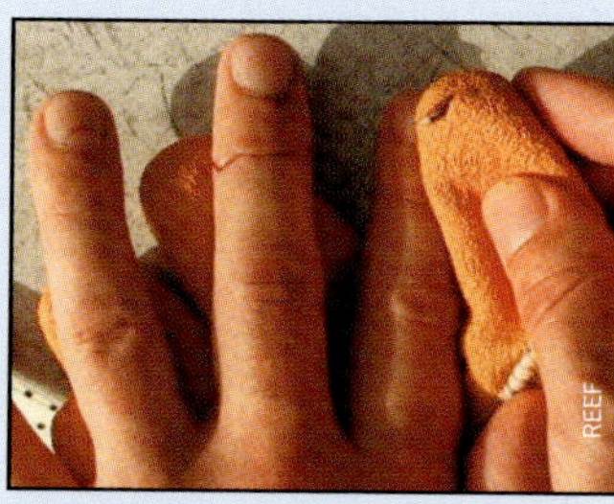

Figure 4.2 This collection diver's finger was nearly severed by a barracuda that had been conditioned by other divers to take speared lionfish. The puncture-proof glove (*top right*) provided minimal protection against barracuda teeth.

Physical Removal

In the absence of natural controlling mechanisms, physical human removal of lionfish can buy time and protect key resources while new technologies are developed and natural controlling mechanisms emerge.

Where and when to find lionfish
Lionfish are generalists when it comes to habitat requirements. They have been found along shallow shorelines to depths over 300 m and in a variety of habitats, including mangroves, hardbottom, canals, artificial structures, loosely organized debris, coral reefs, ledges, sand holes in seagrass, and rocky shorelines. Lionfish will most often be associated with structures that provide cover, but occasionally they can be observed hovering in the open.

During the mid-day hours, lionfish are commonly found resting under ledges or in sheltered bottom areas. They are most active in the low light periods of dusk and dawn. Divers searching for lionfish during mid-day hours should carefully explore overhangs, crevices, and any nook or cranny that could provide lionfish shelter. Swimming over habitat may not be sufficient to detect the majority of sheltered lionfish. Detailed searches — including slow swimming and careful observations — are needed to accurately detect sheltering lionfish. Timing removal events with low light periods around dawn and dusk may facilitate increased detection. Selecting the best equipment for any given situation can improve success (Table 4.1)

Removal Tools

Markers
During the initial stages of the invasions, managers in some areas of the Caribbean used lionfish markers to assist with early detection and rapid response (Figures 4.3 and 4.4). Lionfish can exhibit site fidelity, and marking locations can help minimize search time during subsequent removal efforts. This approach, however, requires planned follow-up actions to remove lionfish and markers. Delayed return to marked sites may provide enough time for lionfish to move to another location. Over time, abandoned markers can accumulate in the environment, decreasing aesthetics and harming marine life.

Nets
Lionfish removals by divers using hand nets have become the standard for many locations, especially in areas that restrict spearfishing. Nets are also more effective for fish that are too small to spear efficiently. Frequently used nets include those used by professional marine life collectors — typically deep-bellied nets constructed with an aluminum frame and clear vinyl netting (Figure 4.5) or an acrylic frame with monofilament netting (Figure 4.6) — or shallow-bellied mesh bait nets (Figure 4.7).

Since most divers and fishers are not experienced marine life collectors, significant training is often needed in order to teach safe and efficient lionfish capture techniques, including proper net speed, positioning, movements, and handling. Net collection of lionfish may be accomplished solo or in pairs, with each diver using two nets (Figure 4.8).

Table 4.1 Pros and cons of various gear types for collecting lionfish.

Gear type	User	Cost	Benefits	Limitations	Notes
Vinyl hand net	Divers and marine life collectors	Initial expense high	Highly effective for naïve fish and smaller fish, when used properly does not scare fish.	Bulky to carry, fish must be accessible, successful use takes practice, slower captures.	
Mesh hand net	Divers and marine life collectors	Initial expense high	Effective for naïve and smaller fish, when used properly does not scare fish.	Entanglement common, takes practice, slower captures.	
Plastic container (mask box, plastic bag)	Divers	Low	Commonly available.	Places user in close proximity to spines, only useful for very small fish.	
Polespear	Divers and spearfishers	Low	Effective for larger fish, rapid captures, keeps fish at distance from collector.	Misses are common, potential damage to resource, potential abuse, too large for most smaller fish.	Spearing devices may be prohibited or restricted in use to freediving only.
Hawaiian sling	Divers and spearfishers	Low	Effective for larger fish, rapid captures.	Spear leaves direct control of collector, misses are common, potential damage to resource, potential abuse, fish may slide toward collector, too large for most smaller fish.	Spearing devices may be prohibited or restricted in use to freediving only.
Speargun	Divers and spearfishers	High	Powerful enough for the largest fish.	Spear leaves direct control of collector, overkill for most lionfish, reloading and removal of fish is lengthy process.	Spearing devices may be prohibited or restricted in use to freediving only.
Specialty spearing devices	Divers and spearfishers	Varies, often high	Usually designed for ease of use, small enough to carry.	Depending on the device: may require two hands, may be shorter than traditional devices, may not secure fish following spearing, may require specialty parts for repair.	Specialty devices vary in their design and application. Spearing devices may be prohibited or restricted to freediving only.
Traps	Commercial fishers	Low	Already in use for other targeted species.	Bycatch is high.	Lionfish have been reported as common bycatch in some lobster and fish traps. However, no lionfish specific trap has yet been developed.
Hook and line	Commercial and recreational fishers	Low	Already in use for other targeted species.	Reported to be effective only in deepwater (300–600 ft), bycatch may be high.	Squid as cut-bait has been reported as successful lionfish bait.

Figure 4.3 Lionfish marker made with a metal washer, flagging tape, and cork.

Figure 4.4 Lionfish marker deployed on the reef to mark sighting.

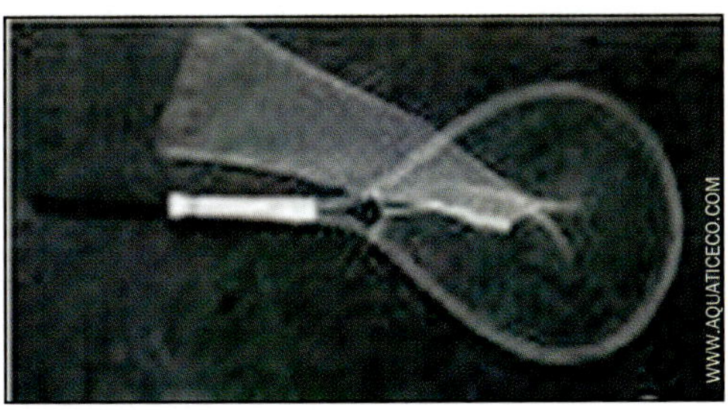

Figure 4.6 Clear mesh net used for capturing lionfish.

Lionfish Marker Programs

The Bonaire Marine Park and the Caribbean Oceanic and Restoration and Education Foundation (CORE) in St Croix, U.S. Virgin Islands, have both implemented programs to manufacture and furnish divers with lionfish markers (Figure 4.3). These markers are made of a metal washer (the anchor) and a strip of surveyor's plastic flagging tape, held suspended from the anchor by a cork.

When lionfish are sighted, the markers are deployed (Figure 4.4) and sighting information is reported to the appropriate managing organization.

While the marking of lionfish may be helpful in the early stages of local invasions, the manpower and management of recovery may soon overwhelm capacity. Some divers in areas using markers have complained of the lack of response to marked fish and the negative aesthetics associated with seeing a dozen lionfish markers on a single dive site.

Figure 4.5 Vinyl hand nets can be highly effective tools for removal, especially for smaller lionfish and in areas where spearfishing is prohibited.

The initial step in netting lionfish is to plan the capture strategy. First, assess the habitat and potential escape routes. Next, determine net placement to prevent movement of the fish to an inaccessible area of the reef. Once the strategy has been decided upon and net placement communicated among team members, slowly position the nets to block potential escape routes. Then maneuver nets *slowly* to surround the fish until it either swims into a net or is trapped between the nets.

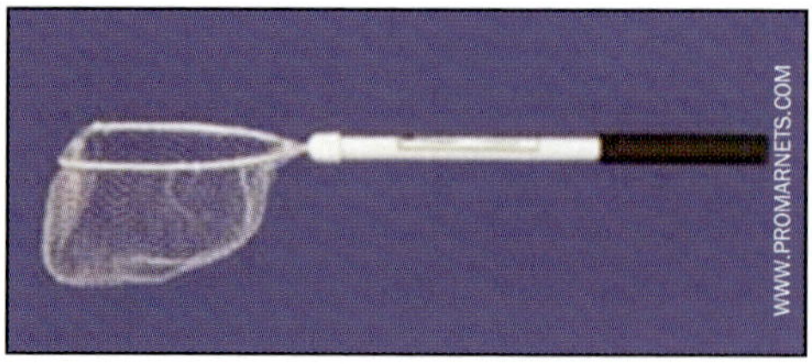

Figure 4.7 Smaller mesh bait nets can be cost-effective and provide access to smaller spaces.

Always avoid chasing, poking, prodding, or scaring the target fish into the net. Lionfish are adept learners and, once harassed, they exhibit avoidance to divers. If target fish move away from nets, it is better to pause the collection effort and allow the fish to re-settle. Unsuccessful lionfish captures in which the fish is harassed extensively can make subsequent collection attempts much more difficult.

Figure 4.8 Divers with nets work in a team to capture lionfish.

After successful capture, transfer the lionfish to either an appropriate underwater storage container (deep captures) or to a storage container at the surface (shallow captures). Care should be taken to exercise proper buoyancy control during captures and to avoid damage to sensitive marine life or habitats.

Spears

Spearing lionfish may be the most effective method for removing large numbers in the least amount of time (Figure 4.9). Lionfish are often sedentary and bold, allowing a close approach with a spearing device. It is important to remember, however, that lionfish are stalking-and-ambush predators and have quick escape reflexes. Even at a very close distance, experienced spearfishers often miss lionfish because of their quick movements. Smaller lionfish can be more difficult to spear than larger lionfish given their smaller target size and ability to seek shelter in small crevices. Despite these challenges, spearing can be a fast and effective method of removal for lionfish in confined spaces that may not be conducive to net capture.

Figure 4.9 Diver with lionfish captured using Hawaiian sling.

As with other removal methods, training should be provided to novice collectors to ensure the greatest safety and success. When learning to use spearing devices, divers should take care to emphasize safety over success. Special training courses have been developed to teach safe spearfishing techniques.

Specific lessons include:

i) Safety first, considering other divers and marine structures and organisms.
ii) Approach target fish slowly and as closely as possible, without causing the fish to move.
iii) Avoid shooting a lionfish that is facing away from the diver.
iv) Aim for the area immediately behind the head of the fish, as lionfish heads are bony and difficult to penetrate with a spear.
v) Exercise careful handling and positioning of the fish after spearing. Lionfish are seldom incapacitated by a single spearing event and may often escape from spears held in open water.
vi) When swimming in open water or handing speared fish up to a vessel, be aware of opportunities for the fish to either flip off the spear or slide down the spear toward the diver.
vii) Be aware of potential predators attracted by spearing activity and do not decapitate or gut lionfish in the immediate area, which may attract unwanted predators.
viii) Obey all spearfishing rules and regulations, including area restrictions and gear types.

Containment

Many different types of containment are being used in lionfish removal efforts, and a few key elements should be considered when choosing which containment type is best suited for the task. Safety should be the priority. Avoid any container that allows spines to freely penetrate or extend beyond its walls. Containment equipment should provide ease of introducing fish and a secure method for storing them. During spearfishing activities, divers may also want to consider the ability of containers to retain the blood and scents of speared fish to avoid encounters with predators. Fish should be kept in whole condition (with the exception of optional spine clipping) to reduce blood and fish particles in the water and to facilitate data collection on the captured specimens.

Commonly used storage containers include clear dry bags (Figure 4.11) and buckets (Figure 4.12) or tubes with perforated lids (Figure 4.13). Dry bags are effective at containing the scent of speared fish and the clear vinyl material offers excellent visibility when transferring additional fish to the bag. Lionfish spines are capable of penetrating vinyl, although this occurrence is infrequent. Gentle handling of a bag filled with lionfish reduces the risk of spine penetration through the vinyl.

Bucket- and tube-type containers can be constructed of commonly found supplies, including PVC pipes or hard plastic buckets with secure lids. Simple star pattern cuts along the bottom or lid of a bucket provide a trap door and the ability to easily remove the lionfish from the spear. Specialty lids can also be constructed of flexible materials, such as thick neoprene or rubber, to prevent deformation of entry points.

Clipping Spines

Some divers clip lionfish spines immediately after capture to reduce the potential of sting encounters. Others simply introduce the captured lionfish directly into a storage container. Clipping of spines is not difficult, but it does introduce additional handling of the fish and may not be approved by some governments or universities' animal-welfare committees.

To clip spines, hold the fish securely by the head and use a pair of sea snips or heavy kitchen shears to sever the spines close to the base (Figure 4.10). Dorsal spines are best cut by starting at the rear of the fish and working forward to prevent the spines from lying flat along the back. Clipping of pelvic and anal fins needs only to include the first spine of each pelvic fin and the first three of the anal fin (the rest are rays and are not venomous). It is not necessary to clip tail or pectoral fins.

Figure 4.10 Lionfish with dorsal spines removed.

Figure 4.11 Lionfish in clear dry bag.

Figure 4.12 Lionfish collection container made using a plastic bucket.

Some dive operators have chosen to include lionfish collecting in their regular dive activities by indirectly involving their customers. Following a detailed briefing on the boat, divers are encouraged to search for lionfish during their dive and notify the in-water divemasters upon locating a fish. Divers are given opportunities to photograph and watch the fish, and then the divemasters collect the fish using hand nets. Following the collection the fish is placed in a clear dry bag, which can then be passed around underwater for the divers to closely observe the fish and engage in photo opportunities. Rather than excluding their clients from lionfish collections, these operators engage their divers and often receive comments that the collection was the highlight of their diving trip.

Steps for Dry Bag Captures

When introducing lionfish into dry bags, divers following these key steps will minimize escape of captured fish and risk of stings:

i) Place the bag on the bottom as mid-water transfers often allow lionfish to escape,
ii) Hold the fish securely (with protective, puncture-resistant gloves) by the head,
iii) Introduce the fish as far into the bag as possible,
iv) Close the bag securely around the arm of the diver holding the fish,
v) After communicating all is ready, let go of the fish,
vi) As the fish moves to the bottom of the bag, block potential escape by placing a protected arm or device (spear, net, etc.) across the bag to prevent the lionfish from moving toward the opening,
vii) Remove the diver's arm from the bag, and
viii) Roll the bag closed and secure by clipping.

Alternative removal equipment
While many lionfish removal devices are commercially manufactured and available on-line or through local retailers, inventiveness and creativity in local and personal devices should not be dismissed. Often, less expensive and user customized equipment can be assembled using locally found parts and materials and creativity. Consideration for safety and effectiveness should always be exercised in the manufacture of personal devices.

Figure 4.13 Lionfish collection using a customized PVC tube.

Often, commercially available designs, which have been proven in their effectiveness, can be modified to incorporate novel uses of readily available and inexpensive local materials (Figure 4.14). Additionally, many new tools and devices will appear on the market and in personal removal toolkits that may prove more effective and less costly than those currently available.

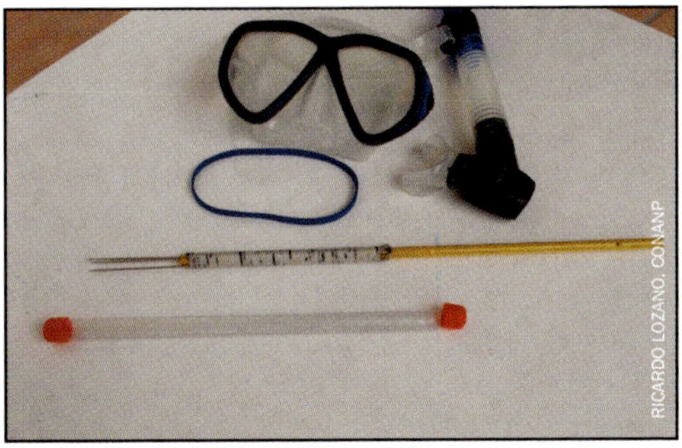

Figure 4.14 Locally made and relatively inexpensive lionfish spearing device.

In Cozumel, for example, creative divers have developed their own locally designed version of polespears, using readily available and inexpensive materials. Spear shafts are made from discarded PVC or fiberglass rods; rubber bands are recycled or made from punctured bicycle innertubes; and paralyzer tips are fashioned from broken bicycle spokes or wire from discarded boat steering cables.

In some instances, collection activities may be restricted to snorkeling and freediving owing to obstacles such as regulatory gear restrictions or lack of available equipment. Collection of lionfish while freediving can be a very effective means of removal in certain conditions. While the majority of collecting tools and techniques are similar to those used with scuba, a few additional considerations are worth noting. First, surface support can allow collected fish to be handed over directly to shore or surface vessel personnel for containment, eliminating the need for in-water storage containers. Fish are simply passed up to surface support in the nets or on the spear and deposited in the on-board storage container. In situations where surface support is not readily available, divers should take care when placing fish into inwater containment, especially noting the ability of the fish to escape through the open entrance of a containment device (Figure 4.15). Additional safety measures and considerations should be addressed during freediving activities, such as increased pace of individual capture, freediving skill level of collectors, and adherence to proper freediving safety procedures.

Traps

While spearing and netting lionfish are active removal techniques, trapping provides a passive method. In areas with trap fisheries, lionfish bycatch frequently occurs, especially at deeper depths. To date, no successful "lionfish only" traps have been developed, and bycatch of native species is high in relation to numbers of lionfish captured. Future gear development efforts could result in more effective trap designs, baiting types, and deployment schemes.

Figure 4.15 Snorkelers contain lionfish in nets before transferring the fish into a storage device at the surface.

Hook and line

Lionfish stalk and consume their prey whole and feed on a wide variety of fish and crustaceans. Even aquarists undergo considerable effort to acclimate captive lionfish to take dead food items. Hook-and-line catches of lionfish on shallow reefs are rare; however, there are some reports of incidental captures in deepwater (150–200 m) snapper and tilefish fisheries and infrequent catches in shallow water by recreational fishers (Figure 4.16).

It is unknown what factors influence these captures, but it is possible that hook-and-line catch from deepwater can be a removal tool. Further information is needed to quantify the effectiveness of lionfish bycatch for local control.

Figure 4.16 Angler with lionfish captured on hook and line.

Safety

Success in capture should never take precedence over safety. One of the possible reasons lionfish have become so invasive in Atlantic waters could be related to their venomous defense. Venomous spines provide a powerful deterrent to predation and can also provide a significant sting to would-be collectors. Puncture-resistant gloves are one of the most important tools in a lionfish collector's toolkit (Figures 4.17 and 4.18). In contrast to standard working gloves, or even Kevlar™ gloves, which are designed to protect against cuts, specially designed and manufactured puncture-resistant gloves provide an added measure of protection against extremely sharp lionfish spines. While not 100% foolproof, use of these gloves will minimize stings and provide safer handling of fish during collection, handling, and in preparing the fish for end-use.

Figure 4.17 HexArmor® puncture-resistant gloves.

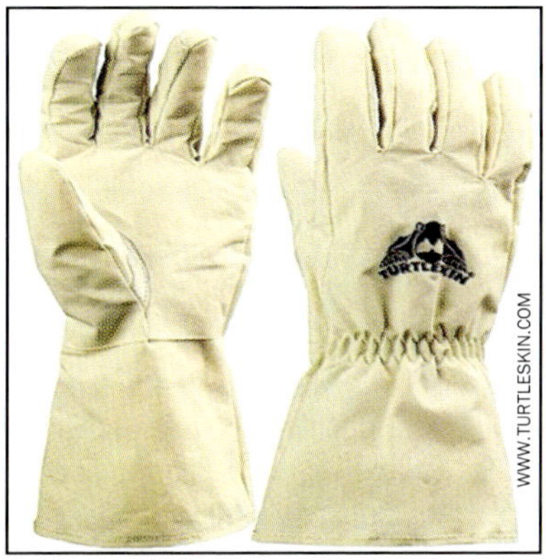

Figure 4.18 TurtleSkin® puncture-resistant gloves.

Safe diving and storage practices should also be incorporated into local lionfish collecting protocols. There are no documented cases of deaths attributed to lionfish envenomation, however, a rapid ascent during diving as a result of a lionfish sting could prove debilitating or fatal, if undertaken as a result of panic following a sting. Improper storage of lionfish on the collecting vessel or in the lab could also provide the opportunity for severe stings. Storing fish in a well-marked and puncture-proof container can help reduce these unnecessary risks.

Envenomation and first aid

A lionfish sting can be a painful and scary experience. For most stings, simple and rapid first aid treatment can be very effective at relieving pain and reducing effects of the venom. Since the venom is a proteinaceous neurotoxin, heat will denature the venom quickly. First aid includes application of heat as hot as one can stand without scalding. The sooner a sting victim receives first aid, the better, as this treatment relieves the pain and begins to denature the venom, possibly reducing more severe effects or complications. Lionfish spines are very sharp, but are not brittle like urchin spines. They tend to enter and exit very quickly, and it is rare for spines to break off. As with any wound, the area should be cleaned, checked for possible debris, and treated by a qualified medical professional.

Treating a Lionfish Sting: Sources of Heat in the Field

Lionfish venom is a proteinaceous neurotoxin, and heat will denature the venom quickly. The key to suppressing the sting is to find safe sources of heat, such as those listed here. Remember not to scald! A burn may be worse than the sting itself. Since numbness can be a sting symptom, be sure to test water temperature with an unaffected part of the body to prevent scalding.

- *i) Bring a thermos of hot water when collecting lionfish.*
- *ii) Nearby restaurants, marinas, or residences can usually heat water in an emergency.*
- *iii) Vessel exhaust water or engine coolant system may be a source of heat.*
- *iv) Lay a wet shirt or towel over an engine block for heating, then wrap.*
- *v) Heat packs do not normally reach high enough temperatures to be effective, however, the reusable packs may provide relief for short periods (5–10 minutes).*
- *vi) Urine is body temperature and is not effective.*

Locations and Frequencies of Removals

Prioritizing locations for lionfish removal

When resources are limited, the most effective method of maximizing results is to prioritize allocation of these resources. This can be done by assessing costs and benefits of removals across locations and utilizing multiple removal methods and personnel. Determining priority areas for removals involves consideration of many viewpoints from various stakeholders. Priorities may differ depending on the perspective. Fiscal and human resources also come into play when determining removal priorities, as do logistics of removal methods. While one area might be deemed most important for conservation reasons, it could be expensive or logistically difficult to engage in removals at that site. Likewise, it might take little effort or expense (through volunteer efforts, for example) to remove lionfish from an area, even though it is ranked lower in ecological importance.

Balancing removal effectiveness with logistical considerations and costs is a key function of management decisions. Many of the examples presented below are not mutually exclusive. There could be some sites that rank high across multiple ranking criteria. Weighing these factors should be a significant component of determining removal priorities.

Marine protected areas

Marine protected areas (MPAs) or reserves are designed for the preservation of marine resources and to minimize effects of extractive activities. For this reason, most have prohibitions on collecting or removing any resources from within their boundaries. Unfortunately, the boundaries are not heeded by lionfish and many MPAs, owing to their protected nature, can become sanctuaries for lionfish. In order to protect the biodiversity and natural state of these special areas, unique provisions are often needed to facilitate lionfish removals. Many resource managers are sensitive to the need for lionfish control, but may be hesitant to allow open access or removals in these areas. While the potential for regulatory abuses exists, preventing removal efforts may prove more damaging than potential abuses in a well-managed effort. Programs to license or permit key stakeholders have been effective and, when implemented under formal programs, they have resulted in low levels of abuse.

Conversely, lionfish management programs that are designed and authorized to control lionfish specifically within MPA boundaries may find decreased success because of minimal removals outside the MPA boundaries. In fact, successful removal programs will take into account the connectivity between managed and unmanaged areas.

LOBSTER OR LIONFISH? NEW OPTIONS FOR DIVERS

In many areas, divers are now engaging in lionfish removals as part of their regular diving activities. Rather than targeting lobster or grouper, these divers are choosing to spend their dive time searching for and removing lionfish. Many dive operators are also encouraging staff to remove lionfish that are sighted during their dive and allowing the staff to keep the fish for later consumption.

In Bermuda, divers are permitted to remove lionfish from areas traditionally off-limits, using tools and equipment traditionally disallowed. They are required to attend a lionfish workshop and to fly a special lionfish removal flag (Figure 4.19), when they are collecting lionfish. In Puerto Rico, resource agency staff and appropriate researchers are issued collapsible polespears and authorized to remove lionfish that are incidentally encountered during other dive-related activities, such as mooring buoy maintenance, research, or monitoring.

Figure 4.19 Bermuda lionfish flag issued along with a special license to spear lionfish on scuba. The flag informs enforcement officers that the required license has been obtained.

Tourism areas

The likelihood of beachgoers and swimmers encountering lionfish is low. Lionfish prefer habitats with structure(s) and are most often inactive during the day. Even with low risk of envenomation, these public areas may warrant removals due to their importance to tourism and economies. In addition to these easily accessed areas, the region derives much of its tourism through diving and snorkeling – activities that place visitors in closer proximity to preferred lionfish habitats. Priority tourism areas may rank high for removals because of the potential of encounters with lionfish and the desire to maintain visibly healthy marine populations as tourism attractions. Lionfish removals from these areas also provide increased opportunities for education and awareness of tourists and the public, as collection activities occur.

Areas of ecological importance

MPAs usually incorporate some habitat diversity, yet there may be other areas that are ecologically important to the health of local marine systems. Mangroves, seagrass beds, and hardbottom areas often serve as nurseries for juvenile fish and invertebrates that are key to maintaining future stocks of adults. Lionfish are known to inhabit all of these areas, and some priority may fall on protecting these sites from predation. Other ecologically important areas — including spawning sites and cleaning stations — may also be vulnerable to lionfish impacts and therefore considered during prioritizing.

Areas of high lionfish colonization

Areas with high lionfish densities can cause significant declines in prey communities and may displace other marine life. In addition, these sites may serve as areas of enhanced lionfish spawning activity and potential recruitment centers for other lionfish. Predation in these high-density areas can surpass the ability of native prey to recover and may cause more significant decreases than predation spread over a larger area. Areas known to harbor large densities of lionfish may be high on the list for targeted removals.

Areas of low lionfish re-colonization

Areas determined to have slow rates of re-colonization following lionfish removal may be priority removal areas due to their high success rates in maintaining low lionfish abundances and minimizing impacts over a great time period. Prioritizing these sites may be a cost-effective approach, as they may need to be revisited less frequently than sites with high re-colonization rates.

Easy removal areas

Some areas that are easily accessed and cleared of lionfish may make a priority list simply because of the ease of removal. Sites frequented by divers willing to adopt a reef, sites nearshore and easily accessed, and sites that allow for effective removals with infrequent visits may all be worth considering for their effectiveness in training, maintaining morale, and reducing lionfish movement to other, possibly inaccessible, areas.

Removal frequency

The frequency of removal needed to control lionfish can vary widely depending on many factors, including the target density, stage of the invasion, habitat, and recruitment pressure. Human and fiscal resources will limit typical removal programs, and the most effective removal frequencies will consider a combination of available resources, removal priorities, and effectiveness. While removal frequencies may vary, it has been shown that active removals on a local scale can significantly reduce and maintain low lionfish densities.

As examples, divers in the Bonaire Marine Park and in the Florida Keys have been removing lionfish from commonly dived sites since the invasion began in those regions. Anecdotal information indicates the densities and sizes of lionfish in those heavily removed sites are much less than in surrounding un-dived sites. Additional research being conducted in the Bahamas is showing regular removals on shallow patch reefs to be highly effective at keeping densities far below control sites.

Lionfish Removal Events

One of the most effective methods for engaging the stakeholders — including the public — in lionfish removals includes organized events and removal programs. These efforts often contain a social component and opportunities for training, increasing awareness, and spreading information through media. In addition, depending on the type of event, they can often remove large numbers of lionfish in a short period of time.

Derbies

Fishing and spearfishing tournaments have long been held for commonly targeted species. Recently, derbies have begun to be organized specifically for lionfish. Successful derbies typically include educational components that cover basic biology, ecology, impacts, collecting and handling techniques, and details of rules and regulations associated with lionfish removals. Often offering significant prize money through sponsorships, the derbies also incorporate data collection for science and lionfish-tasting opportunities for the general public. Single-day derbies have been known to bring in thousands of fish from small geographic areas and may have measurable effects on local lionfish populations (Figure 4.20).

Monthly contests

To encourage ongoing removals throughout the year, groups in a number of locations have implemented monthly or seasonal contests. Media attention and outreach opportunities might not be as productive as for single-day events, but the

Derbies Remove Lionfish by the Thousands

The Green Turtle Cay Lionfish Derby held in June 2009 in Abaco, Bahamas, was the first lionfish derby held in the invaded region. In a single day, 1,408 lionfish were removed from local waters by 18 teams competing for prize money and bragging rights. Prizes are offered for most, largest, and smallest lionfish. Since that time, numerous derbies have been organized around the region, serving to educate divers, raise awareness, and remove lionfish. The Florida Keys hosts a derby series (Figure 4.21) and some recent derbies in Mexico and the Bahamas have removed well over 2,000 lionfish in a single day. Monthly ongoing removal contests are also a successful way both to encourage ongoing removals and offer small prizes and recognition to divers and dive operators.

Visit http://www.lionfishderby.com for derby examples.

Figure 4.20 Lionfish derbies can provide intensive local removal success as well as excellent opportunities for outreach and the chance for participants to feel a sense of community and purpose.

collective results of ongoing removals could be of greater significance.

Adopt-a-reef programs
Many groups are interested in helping to minimize lionfish impacts on an ongoing basis. Working with dive clubs, NGOs, and other organized groups to pick key sites for targeted removal efforts can be very effective in keeping lionfish numbers reduced in those areas. Often, groups can be self-organizing and require minimal management, though follow-up, recognition, and periodic supervision will help maintain ongoing morale and effective removal programs. It is important to design programs with appropriate levels of scale and realistic expectations of effort in order to avoid overtaxing abilities.

Figure 4.21 Florida Keys derby poster.

Organized efforts by divers
Organized and directed removal efforts have been undertaken in many locations. Divers in the Florida Keys have been permitted to remove lionfish from marine protected areas and they keep numbers down through ongoing removals during regular visits to these sites. The Caribbean Lionfish Response program in the U.S. Virgin Islands organizes regular removal dives to preselected sites. Divers with Ecotono in Puerto Rico have adopted certain sites for regular lionfish visits and removals. Many dive operators throughout the region remove lionfish from their sites on daily visits to keep numbers down.

End-use and Disposal

Resource managers put much effort into designing and implementing removal plans and programs; however, the end-use of the removed fish is often overlooked. Important information can be garnered from collected fish. (Further, safety and conservation issues may come into play when considering the end-use of removed lionfish.) Simply recording information on size of the fish collected can be useful in determining removal effectiveness and impacts.

Removed fish can also be dissected and key data gathered to aid in understanding their biology and ecology. Not every fish needs to be fully dissected, but developing a plan and protocol for gathering information as a part of removal efforts can be important (see euthanasia methods described in Green et al. 2012). Some removal efforts may also provide a source of income through sales of fish for local consumption. Allowable and encouraged end-uses should be included as part of lionfish control and removal plans.

Final disposal of collected fish should also be addressed. Venom in lionfish spines denatures rapidly with heat, but the sharp spines can inflict serious wounds even without venom. Guidelines for discarding fish and/or their spines should include provisions to reduce potential interactions and risk to other humans or animals. Puncture proof bottles or jars with caps can provide relatively safe and inexpensive disposal containers. Some recent removal efforts have included the feeding of lionfish (immediately following their collection) to predatory marine life, including sharks, groupers, snappers, and eels, but consideration should be given to potential risks of spine injuries to these predators as well as behavioral changes due to association of feeding with humans.

Is Feeding Lionfish to Top Predators a Good Idea?

The following opinion is an excerpt from http://myfwc.com/research/about/outreach/science-behind-management/marine-fish-feeding/

"In 2001, the Florida Fish and Wildlife Conservation Commission (FWC) unanimously voted to prohibit divers from feeding marine life in Florida. You might be wondering why the FWC believes these 'interactive marine experiences' are harmful. After all, the fish get an easy meal, and divers get entertained, so what's the problem?

"Overall, feeding marine fish is a bad idea for everyone, including divers, fish and the ecosystem. Contrary to popular belief, fish have memories and can learn. Through behavioural conditioning, fed animals learn to associate people with a meal. When this happens, fish anticipate the hand-feeding experience and depend on handouts from divers . . .

"Hand-feeding-induced attacks on humans do occur. Feeding wildlife can place people in harm's way . . .

"Moray eels, sharks, barracuda, groupers and a host of other species can pose an increased danger to divers as a result of hand-feeding."

TRAINING

Fishers and many divers are accustomed to removing and handling Caribbean fish species as part of their every-day work. Lionfish, however, are unlike any other fish in the Caribbean. Their behavior and long slender dorsal spines can prove to be challenging for removals by even experienced fishers and divers. In addition, much misinformation exists in the popular culture and media about the lionfish invasion and the biology and ecology of the fish. For these reasons, implementation of well-designed and presented training programs for lionfish collecting and handling are extremely useful (Figures 4.22 and 4.23).

One of the key elements in any training program is the safety of those doing the collecting. Proper techniques and hands-on demonstrations should be included in all removal presentations, and participants should have opportunities to handle lionfish under supervision whenever possible. Specific guidance in locating venomous spines and proper positioning of gloved hands to minimize stings should be given. Additional information on first aid for lionfish stings should also be covered, with recommendations for field applications of heat as first aid.

Figure 4.22 Lionfish workshop utilizing the classroom.

Figure 4.23 Lionfish workshop in the field.

Courses

As the lionfish invasion progresses, numerous courses and training programs will be sure to follow. It is important for experienced reviewers to vet these training programs to ensure that safety protocols are included and all information is accurate. Program managers may want to endorse specific courses or removal programs (once they have been properly reviewed) to encourage passing along of accurate and timely information; however, care should be taken to maintain proper guidance and updates of information.

Materials

Many materials for conducting removal training can be ordered or obtained via the Internet. It is anticipated that specific training videos and slide presentations will follow this manual and be made available free of charge for use by marine resource managers and their key partners.

Liability

Any time programs deal with potentially harmful animals, questions regarding liability issues may arise. Many countries have developed lionfish-specific statements of understanding or liability releases to reduce the potential legal issues associated with injury. It is wise to consult with local legal authorities before undertaking lionfish removals to determine the level of liability and methods to address or minimize legal risk. All participants in any removal effort should be fully briefed on safety issues, potential risk, and first aid; and they should participate fully of their own will. No person should ever be forced into a position that they are uncomfortable with in handling either dead or live lionfish. (See Appendix 3 for example of liability release.)

Training strategies

Some specific training strategies include:

i) Conduct in-water training to motivate volunteers and enhance effective control,
ii) Facilitate "Train the Trainers" programs,
iii) Provide training on how to construct removal gear,
iv) Provide training on first aid and medical treatment,
v) Provide cooking classes and training on lionfish consumption,
vi) Collaborate with the dive industry and fishers,

vii) Establish protocols for proper use of specimens following collection, and
viii) Use available online technology and social media tools.

REMOVAL INCENTIVES

The incentive for participating in lionfish removals varies among individuals, and often depends on whether their livelihoods or recreational interests are connected with the marine environment. Incentives can vary widely depending on the interest and motivation of the individual. (Even those who make their living in or on the water often need an incentive.) Some may want to remove the fish because they want to protect native marine life and ecosystems, some may want to remove fish for food, and some to protect their livelihood or for economic benefit. It is important to recognize which incentives motivate individuals to remove lionfish and to foster those motivations.

Reasons for Stakeholder Involvement

Divers
The desire to protect marine resources is a key motivator for many recreational divers, tourists, and tour operators. Their livelihoods and recreation depend on healthy marine systems, and many understand that removing lionfish whenever possible will help to protect those resources. For these stakeholders, recognition of their efforts and feedback on results often provide the motivation and incentive needed to maintain removal activities.

Fishers
Many fishers rely on their catch to provide a substantial portion of the diet for their families. Eating their catch can also be a strong incentive for recreational fishers to fish. Promoting the removal and consumption of lionfish to these stakeholders can help provide incentive for their targeted removal. Many subsistence fishers are already considering lionfish as worthy of keeping and consuming when caught.

Commercial fishers and divers, including spearfishers, spend their time capturing fish to make their living. Specific fish species are often targeted for their high dollar value and marketability. Developing consumer demand for lionfish can help elevate the dollar value of the fish in the market and provide incentive for commercial removals. Because of the significant effort required in removing lionfish and their small size as compared to some other commercially targeted species, prices paid will likely need to be higher to compensate for the time required in harvesting. Luckily, lionfish flesh is deemed high quality and has been shown to be better in omega fat content than many commonly consumed native species.

Bounty programs
Bounty programs have been attempted in a few locations and have been short, lived, with little promise for success. In addition to limited funds, bounty programs lack provisions for developing stewardship ethics and are typically for the short-term monetary benefit of a few. Once funding runs out, removal effort is often discontinued and the invasion continues to progress unabated. Additionally, as markets develop for lionfish, bounty programs position governments in direct competition with private enterprise, possibly hindering more than helping.

Long-term nature of the issue
In any incentive program, care should be taken not to create expectations of short-term eradication. The region will likely be facing the lionfish issue into the foreseeable future and hinting at eradication or short-term solutions will only develop resentment as the problem continues. Realistic expectations of local control and minimizing impacts through long-term incentive programs are key in developing and presenting incentive options.

Incentive Sources

Donations
Incentives do not always have to come in the form of money. In fact, many incentive programs work together with private enterprise to offer donated items or services of value to lionfish collectors. Monthly contests, dive-tank air for fish exchanges, and raffle drawings have all been used as ways to include donated goods as incentive for removals.

Markets
Ongoing programs based on the commercial use of the lionfish may be the most sustainable model for incentives. Market-based payment for food fish is already in place for most coastal communities and simply developing the supply-and-demand chains can provide ongoing payments for fishers to target and remove lionfish. Markets for smaller fish to be used as aquarium fish encourage removals of juveniles and many other market-based uses (such as novelties and jewelry) are being explored. Managers may also consider the timeline of the invasion in determining when to begin introducing the concept and promotion of lionfish consumption. Recently invaded areas may even consider priming the market with pre-packaged lionfish imports as a method to enhance consumption prior to the availability of locally caught lionfish.

Subsidies
While bounties tend not to work very well, subsidies may. Providing help in the form of price control or subsidy, shipping costs for exports, or specific collecting tools, training, and equipment may encourage removals and increase market supply. Governments often subsidize agricultural products, and following proven examples of effective subsidy programs may be useful.

Recognition
Sometimes the only incentive required is a little recognition for efforts well undertaken. Recognition is especially important for volunteers, but should not be overlooked in working with other stakeholders or sectors of control programs. Recognizing participation in removal training events, removal efforts, and support is often as simple as issuing short media releases on a regular basis. Additional recognition through certificates, pins, hats, or shirts can go a long way in encouraging participation and continued removals. Never underestimate the effect of public recognition in encouraging participation.

Specific Strategies for Incentives

Specific strategies and issues related to incentives for lionfish capture include:

Commercial incentives

i) Identify novel uses and alternative products for lionfish (spines, jewelry, biomedical),
ii) Evaluate market use of juveniles for the aquarium trade (pros and cons),
iii) Educate the public about lionfish as a food resource,
iv) Clarify negative perceptions about the dangers of lionfish,
v) Develop links between supply and demand for lionfish,
vi) Recognize potential risks of creating a market for lionfish (i.e., fishers interest to "grow and maintain" lionfish on the reef),
vii) Highlight lionfish consumption as a "green alternative" to other fish species,
viii) Ensure quality control of product (e.g., quality of meat or presence of toxins),
ix) Recruit marketing experts,
x) Be open to creative marketing, yet bring a balanced perspective to limit economic and cultural "acceptance" of an invasive species,
xi) Consider government subsidies (e.g., those similar in agriculture, farm, export, etc.),
xii) Establish a campaign in partnership with the aquarium industry,
xiii) Consider marketing strategies for all sizes of fish,
xiv) Create tax incentives or other government incentives (e.g., dive operator participation, fisher licensing, etc.),
xv) Ensure that the invasive (negative) nature of lionfish in the Caribbean is highlighted in all marketing materials,
xvi) Link decline in lionfish to restoration of native fish populations,
xvii) Investigate market use of bycatch of lionfish.

Public incentives

i) Establish and support tournaments and derbies,
ii) Create adopt-a-reef programs for removal,
iii) Recognize efforts of volunteers and partners,
iv) Provide discounts for purchase of scuba equipment/air fills,
v) Purchase, in whole or in part, removal gear for volunteers (e.g., fishers),
vi) Promote individual food consumption (i.e., "catch your own dinner"),
vii) Include control strategies in management plans, business plans, and reports for groups seeking funding,
viii) Use bounties if/when appropriate,
ix) Provide direct government payments if/when appropriate,
x) Generate prizes and rewards for removal,
xi) Establish and maintain national records (e.g., biggest fish/most fish captured),
xii) Award academic credits to university students for community service,
xiii) Promote lionfish control as a "Green Initiative," and
xiv) Address liability issues.

Monitoring

Monitoring the effectiveness of lionfish removals and setting removal frequency targets is critically important to allocate removal resources effectively. Chapter 5 provides extensive information on developing monitoring methods and efforts – specifically for lionfish.

No control program should be designed without due regard for this assessment and evaluation. Effective monitoring includes not only monitoring for impacts, but also monitoring of removal effort. These records can also be used as positive feedback for removal participants to maintain positive support of, and participation in, removal programs.

Fish Food for Thought

There has been considerable interest throughout the region in developing commercial markets for lionfish as a food fish. One criticism of this effort has included the potential for smaller juvenile lionfish to be left by commercial fishers in favor of larger, market-ready lionfish. While the potential for this is real, one possible mechanism to prevent this is to also create a market for juveniles. While some believe it is counter to combating the invasion, large commercial markets do exist for juvenile lionfish in the aquarium trade.

Puerto Rico currently exports approximately 200–300 juvenile lionfish per week to supply the U.S. aquarium trade. Similarly, Florida Keys marine life collectors routinely remove and sell juvenile specimens. While the aquarium trade is implicated as the initial source of the invasion, the number of fish released was few. Even if a small percentage of collected fish were to end up back in the wild, the numbers removed would be greater and may contribute to a reduction in impacts. Further, there are economic benefits to fishers and collectors for lionfish removals.

In some countries, however, there is concern that allowing aquarium trade in invasive species like lionfish could elicit additional introductions of lionfish or other non-natives. Potential restrictions on the sale and husbandry of non-native marine fishes in coastal zones, marking individual animals, and initiating permitting fees have all been suggested as ways to address possible releases.

CHAPTER 5

Monitoring:
An Essential Action

Stephanie J. Green

This guide outlines a suite of tools and techniques for minimizing the effects of lionfish on invaded marine ecosystems, and their associated economies and societies. Strategies to manage the lionfish invasion focus on controlling the size of local lionfish populations, because the number of lionfish in an area largely dictates the severity of their impact. For resource managers who are designing and implementing control strategies, a key question is: "How successful are my actions?" This chapter provides managers with the tools to answer this question. It outlines a crucial — and often overlooked — component of control plans for lionfish: designing and implementing monitoring programs to evaluate their success.

Measuring success is essential because it:

i) Gives an indication of whether the lionfish control strategy is achieving its goals, and
ii) Provides direction on how actions can be modified to improve control (Figure 5.1).

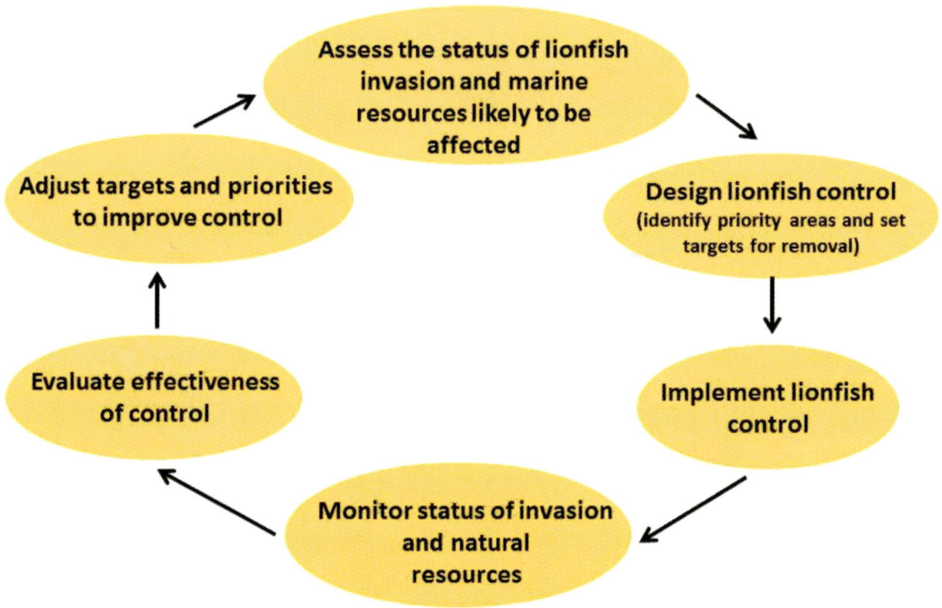

Figure 5.1 Monitoring is an integral component of the Adaptive Management Cycle for lionfish control.

Lionfish are a new stress on invaded marine ecosystems. Likewise, monitoring lionfish populations and their effects – and ultimately the success of control in mitigating them – requires new techniques and approaches. This chapter is divided into four parts. The first describes protocols and tips for assessing local lionfish populations. The second describes the main ecological effects of lionfish and provides monitoring indicators for each effect. The third part describes organismal monitoring, through lionfish dissection; and the fourth considers monitoring socioeconomic impacts related to fishing, tourism, and human health. Suggestions for appropriate monitoring protocols are included throughout the chapter.

A range of programs are underway in the western Atlantic, Caribbean, and Gulf of Mexico to monitor the impacts and management of stressors, such as over-exploitation, habitat degradation, pollution, and human mediated climate change. In many cases, existing programs can be modified to include protocols that assess lionfish and their effects. To aid in integration with existing programs, examples of protocol modifications are provided.

MONITORING LOCAL LIONFISH POPULATIONS

Several key questions managers will want to ask as they begin to develop a monitoring program for local lionfish populations are:

i) How are lionfish numbers changing over time?
ii) Are some locations being colonized more quickly or heavily than others?
iii) How effective is removal in controlling local lionfish numbers?
iv) How much effort is required to control lionfish numbers?

There are several methods that managers can employ to track changes in local lionfish populations over time and in response to management actions. These methods can be broadly divided into *fisheries independent* and *fisheries dependent* data collection.

Fisheries Independent Monitoring

Fisheries independent monitoring activities are those conducted separately from commercial fishing operations. All sectors of society can be mobilized to collect valuable fisheries independent information on local lionfish populations. Table 5.1 describes data collection methods, training requirements, and the applications of data gathered for three main sectors: the public, trained volunteers, and resource managers/researchers. Methods vary in the amount of time and training required, their relative geographic scale, and the level of detail at which data are collected.

Public

The public provides a vast resource of eyes both in and on the water. With minimal training, members of the public can opportunistically collect and report lionfish sightings during recreational activities, such as diving, snorkeling, fishing, and swimming. To ensure data quality, a person's level of training and expertise should be considered. Untrained members of the public should be asked to report only basic information (Table 5.1).

Table 5.1 Protocols and considerations for monitoring local lionfish populations, shown by sector.

Sector	Type of data collection	Data collected	Application	Training required
Public	Opportunistic: • Sightings data collected as part of other in-water activities, such as fishing, diving, snorkeling or swimming	Lionfish sightings information: • Number of lionfish sighted • Sighting location (habitat type coordinates, depth, date, time)	Sightings distribution and frequency by location, over time, and by habitat Spatial and temporal patterns of lionfish colonization (aid in determining priority areas for control)	Ability to accurately identify lionfish Awareness of and familiarity with the reporting system
Trained volunteers	Organized: • Data collected during organized in-water lionfish survey projects or incorporated into ongoing monitoring programs	Roving diver surveys of lionfish abundance* Estimates of lionfish size (total length)	Spatial and temporal patterns in lionfish abundance, body size, and biomass Spatial and temporal patterns of lionfish colonization (aid in determining priority areas for control)	Ability to accurately identify lionfish Ability to accurately estimate lionfish length Training in conducting roving diver surveys
Resource managers/ researchers	Organized: • Data collected during organized in-water lionfish survey projects or incorporated into ongoing monitoring programs	Transect surveys of lionfish density** Estimates of lionfish size (total length)	Spatial and temporal patterns in lionfish density, body size, and biomass Spatial and temporal patterns of lionfish colonization (aid in determining priority areas for control)	Ability to accurately identify lionfish Ability to accurately estimate length of lionfish Training in conducting belt/transect surveys

* The total number of individuals per survey. **The number of individuals per unit area.

While relatively few resources are required to gather a potentially large amount of sightings data, resource managers will have little control over sampling effort (i.e., sighting effort) across space and over time. The majority of sightings will likely come from areas that are accessed frequently, such as beaches, snorkeling and dive sites, and recreational fishing areas.

Resource managers should collect demographic information on the groups reporting the majority of sightings, and comparisons between the numbers of sightings versus captures can be used to direct outreach/awareness efforts.

Trained volunteers
Volunteers can greatly increase the amount of skilled personnel devoted to monitoring local lionfish populations, particularly in areas where resource management staff are limited. After receiving standardized training, volunteers can participate in organized data collection using standardized protocols, either on scuba or snorkel (Table 5.1). By coordinating volunteer activities, resource managers can control sampling design, across both space and time. Volunteer activities may require focused logistic support (e.g., boat logistics and scuba tank fills) or may be integrated into ongoing monitoring programs. In all cases, resource managers will need to work with volunteer program leaders to coordinate sampling design and to obtain and manage data.

Resource managers/researchers
With sufficient training, resource managers and researchers can generate high-resolution data on the spatial and temporal distribution of lionfish. Monitoring activities by this sector differ from those of volunteers largely in the amount of time devoted to training and to conducting the monitoring itself (Table 5.1). While monitoring by resource management and researchers is resource-intensive, requiring dedicated funding for personnel and logistic support, it yields high-quality data against which the utility of public and volunteer data can be compared. It also informs site-specific decisions about lionfish control, which require spatially explicit information on lionfish biomass density.

An integrated approach
Lionfish control programs should include monitoring by all three sectors — the public, trained volunteers, and resource managers/researchers — because they represent information gathered at complementary spatial scales and levels of detail (Figure 5.2). For example, data gathered by the public coarsely approximate the relative rate of lionfish colonization and abundance over time and across a broad geographic area. On the other hand, monitoring by volunteers can quantify the abundance and biomass of lionfish within designated geographic areas. Finally, detailed data collection by resource managers and researchers provides information on local lionfish populations over space and time, in relation to control activities at specific research sites. The locations where the public, volunteers, and resource managers/researchers monitor lionfish are not mutually exclusive. Overlap in the monitoring locations can provide information on the relative utility of information generated by each sector for informing lionfish control (Figure 5.2).

Figure 5.2. Circle size represents the relative geographic scale of monitoring by each group, while color represents the level of detail obtained from the data, with darker shades indicating more detailed information.

Centralizing lionfish data
The USGS Nonindigenous Aquatic Species database is the largest repository for spatially referenced sightings information for lionfish in the western Atlantic. The majority of lionfish sightings are derived from personal communications through the online reporting system. Researchers, resource managers, and members of the public can submit sightings information, including geographic location and habitat information, online at http://nas.er.usgs.gov/sightingreport.aspx. This centralized database provides managers across the region with the most up-to-date lionfish distribution information for both their local area and the region, reduces the chance of duplication or data loss, and eliminates the need to allocate local resources to intensively manage sightings data within each location individually.

Tips for Surveying Lionfish

Lionfish are cryptic and patchily distributed
The cryptic coloration and behavior of lionfish make them particularly challenging to detect underwater (Figure 5.3). Lionfish primarily forage during low-light periods at dawn and dusk; during the day, they are most often found sheltering under structures (Green et al. 2011). Yet the vast majority of recreational and scientific diving and snorkeling activities take place during the day, when lionfish are least visible. On densely populated habitats, lionfish will often be found in close proximity to one another and in association with large structural features (such as coral heads or man-made debris). This means that divers must survey a rather large area in order to capture their distribution accurately.

Standard survey methods have a low probability of detecting lionfish
A recent study on Bahamian coral reefs found that belt transect and stationary visual census (SVC) surveys failed to detect more than half of the lionfish within their boundaries as compared with thorough lionfish-focused searches of the same area. Importantly, lionfish detection during belt transects and SVCs varied significantly with the size of lionfish and reef complexity, with smaller lionfish and high complexity habitats resulting in low detection, indicating that the application of a correction factor to standard visual data is not straightforward (S. Green, Unpublished data).

Suggested standard survey procedure
Results over time suggest that thorough lionfish-focused searchers are most appropriate for characterizing lionfish distribution. While the exact transect area may vary by program and location, resource managers must ensure that surveyors take adequate time to make a thorough search of the habitat for lionfish. The observer should take care to look under all overhangs, in crevices, and in cracks in the substrate, using a dive light as needed. A maximum search rate of 10 m^2 per minute is recommended, with even longer search times in complex habitats.

Figure 5.3 The cryptic coloration and behavior of lionfish make them particularly difficult to detect during day-time visual surveys. Can you spot all four lionfish in this photograph?

Because lionfish are often patchily distributed — and the probability of detecting them using standard survey protocols is low — conducting separate, lionfish-focused surveys within an expanded search area is recommended, rather than relying on abundance and density information from all-species surveys. Lionfish-focused transect surveys of 25 m x 10 m (length x width) are well suited to detect lionfish in patchy, heterogeneous marine habitats. A trained observer swims in an S -shaped pattern (Figure 5.4), taking care to thoroughly search all habitat features. The number and size (total length to the nearest 1 cm) of all lionfish within the survey area are recorded. Only lionfish are recorded during the survey. The minimum search time for each transect is 25 minutes.

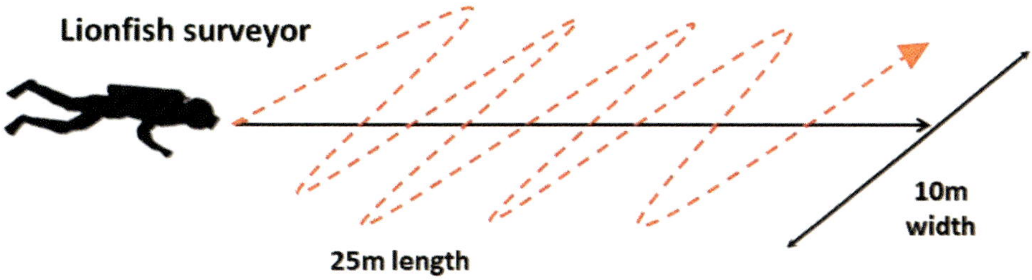

Figure 5.4 A standard approach for monitoring lionfish density. After laying out a 25 m transect tape, a diver makes a thorough, lionfish-focused search by swimming in an S-shaped pattern within the transect area, 5 m on either side of the transect line. The abundance and size (total length to the nearest 1 cm) of all lionfish encountered within the transect area are recorded.

Modifying Existing Volunteer Protocols to Gather Lionfish Data

Volunteer divers and snorkelers currently collect valuable information on marine life through managed programs throughout the region, such as the Reef Environmental Education Foundation (REEF) Volunteer Survey Project. An ongoing effort across the western Atlantic, REEF allows volunteer scuba divers and snorkelers to collect and report information on marine fish populations. These data are used by a variety of resource management agencies and researchers.

REEF volunteers use the roving diver technique (RDT) — a visual survey method well-suited for volunteer data collection (Table 5.2). During RDT surveys, divers swim freely throughout a dive site and record every observed fish species that can be positively identified. Species and approximate abundance scores are recorded on an underwater slate. The goal is to find as many species as possible so divers are encouraged to look under ledges and up in the water column, and to seek out cryptic species. Each recorded species is assigned one of four abundance categories based on how many individuals were seen throughout the dive. The categories are: single (1 individual), few (2–10), many (11–100), and abundant (>100).

REEF volunteer roving diving surveys (as shown here) may be modified to gather data on lionfish abundance and size.

During REEF surveys, divers and snorkelers can also tabulate the size (total length to the nearest 1 cm) and exact number of all lionfish sighted during a 30-minute period. This simple modification to the existing protocol provides useful data on local lionfish populations. For more information, see http://www.reef.org.

Table 5.2 Common underwater visual survey methods used to assess fish communities in the western Atlantic. Methods vary in the number of species surveyed, the resolution of size data collected, and whether density or abundance is obtained. The type of ecological effect monitored determines which method is most appropriate.

Monitoring program	Method	Reference	Metric	Body size recorded?	Set of species surveyed?
Reef Environmental Education Foundation (REEF) fish survey program	Roving diver survey	Schmitt and Sullivan (1996)	Log-scale abundance	No	All species
Atlantic and Gulf Rapid Reef Assessment (AGRRA)	Belt transect survey	Lang et al. (2010)	Density	Binned (5 cm increments)	Commercially exploited and grazers only
Reef Check	Belt transect survey	http://www.reefcheck.org	Density	Binned (5 cm increments)	Selected commercially exploited and grazing fish, and some invertebrates
US Government/ National Parks Service/NOAA National Ocean Service	Belt transect survey	Rogers et al. (1994)	Density	Nearest 1 cm	All species
US Government/ NOAA National Marine Fisheries Service	Stationary visual census	Bohnsack and Bannerot (1986)	Density	Mean, min, max	All species; restricted detection of small-bodied/ cryptic species

Fisheries Dependent Monitoring

Recreational fisheries – derbies and tournaments

Single-day lionfish fishing events and monthly contests are growing in popularity across the Caribbean as a way to raise awareness about the invasion, showcase lionfish as edible, provide samples for research, and reduce local lionfish populations (Figure 5.5). Over time, lionfish derbies that track the catch of lionfish and the effort by participants – as measured by catch-per-unit-effort (CPUE) – can provide an indication of change in local lionfish abundance. Tracking the distribution of lionfish sizes captured at each event can further provide an indication of the effects of fishing on the structure of local lionfish populations over time. Monthly contests can provide similar information on a more frequent basis, though sample sizes may be smaller than in single-day events.

Data to collect at each event:

i) Number of lionfish removed,
ii) Sizes of lionfish captured,
iii) Locations of capture (depth, habitat type), and
iv) Number of participants, gear types used, and time in the water.

Commercial and artisanal fisheries

Information on lionfish bycatch and fisher efforts can provide an indication of how local lionfish abundance and size distribution change over time. The minimum data to record to obtain these estimates includes:

i) Number and sizes of lionfish caught,
ii) Locations of capture (depth, habitat type), and
iii) Number of fishers, gear types used, and time in/on the water.

This information can be gleaned from fisheries observer programs, at landing stations, or by conducting research using gear similar to that employed for commercial purposes.

Figure 5.5. Data collected during lionfish derbies can be used to track local lionfish populations over space and time. During the past four years, recreational fishers have removed more than 4,000 lionfish from waters around the Abaco Islands, Bahamas, as part of the annual Green Turtle Cay Lionfish Derby. Each year, the number and sizes of lionfish caught, the number of participants and gear types used, and the locations in which the teams fished are recorded. (See Appendix 4 for sample data sheets.)

MONITORING ECOLOGICAL IMPACTS

Lionfish occupy coral reef, seagrass, mangrove, estuary, man-made, and deep-sea habitats between depths of 1–300 m. This section focuses on the main ecological effects of their interaction with native species in these habitats: the direct effects of lionfish on their prey and on their competitors, and in-direct effects (mediated by predation and competition) on habitat structure. This section also provides a description of the types of native species likely to be affected by lionfish and the monitoring methods for assessing their populations. Specific instruction on sampling design and implementation of each method can be found in the reference materials cited throughout this section, and summarized in Tables 5.2, 5.3, and 5.5.

Predation

Lionfish consume an array of species and sizes of native fish and invertebrates usually overlooked in standard monitoring protocols (Albins and Hixon 2008, Morris and Akins 2009, Green et al. 2012). These include cryptic and small-bodied species, and the juvenile size classes of larger bodied species (Figure 5.6), which have declined rapidly on some invaded reefs as a result of the lionfish invasion (Green et al. 2012).

Indicators and protocols

To assess lionfish predation impacts, managers can monitor temporal and spatial trends in the diversity, density, and biomass of prey-sized invertebrates and fish, in relation to lionfish density and biomass. The maximum size of prey lionfish can consume is largely determined by their gape size (i.e., mouth size). Lionfish have been documented to consume prey greater than 40% of their total length (TL). As a result, the vulnerability of native species to lionfish predation depends on the size of the lionfish in the area. With lionfish growing to sizes well over 40 cm TL in their introduced range, fishes and crustaceans up to ~15 cm length could be consumed.

For fish prey, visual belt transect surveys recording the identity, number, and size (TL) of all prey-sized fishes are appropriate (Table 5.2). Unfortunately, there are no standard approaches for monitoring invertebrate prey, due to their small size and very cryptic nature.

Suggested standard visual survey procedure for fish prey

Accurately assessing populations of lionfish prey requires a thorough knowledge of Atlantic and Caribbean fish identification (including juvenile life stages, as well as cryptic species such as gobies and blennies) and fish-size estimation training. Replicated transect surveys of 25 m x 2 m (length x width) are well suited to detect small bodied and cryptic fishes on patchy, heterogeneous, marine habitats. A trained observer makes a thorough search of the transect area, taking care to look under all overhangs, crevices, and in cracks in the substrate, using a dive light as needed to look for cryptic fish. The number and size (total length to the nearest 1 cm) of all prey-sized fishes (i.e., less than ~15 cm) within the survey area are recorded. The minimum time for each transect is 30 minutes. Time may be longer depending on habitat complexity and fish density.

PHOTO COURTESY OF NEW WORLD PUBLICATIONS

Figure 5.6. Potential lionfish prey. Lionfish consume an array of native fishes, including small-bodied and cryptic species, such as (top, left to right) the masked goby (*Coryphopterus personatus*) and secretary blenny (*Acanthemblemaria maria*), and juveniles of commercially and ecologically important species, such as (bottom, left to right) the Nassau grouper (*Epinephelus striatus*), and Spanish hogfish (*Bodianus rufus*).

Competition

Lionfish occupy the same habitats and consume similar prey to many species of native fish predators (piscivores and carnivores) and macro-invertebrates (Figure 5.7). Competition with lionfish for food and space may affect the behavior, distribution, growth, survival, and, ultimately, population size of native organisms (Albins and Hixon 2011, Green et al. 2012, Albins 2012). It is possible that competition with lionfish will affect western Atlantic fisheries. For example, research is showing that lionfish grow much faster and consume prey at significantly higher rates than native Coney grouper (*Cephalopholis fulva*), raising fears that they may outcompete this species on invaded reefs (Albins 2012).

Indicators and protocols

Monitoring temporal and spatial trends in the diversity, density, and biomass of large-bodied macro-invertebrates and predatory fish (i.e., >20 cm TL), in relation to lionfish density and biomass, will provide insights into the effects of competition with lionfish. Habitat-scale data are needed to determine if competitive interactions detected during small-scale experiments affect predator populations across invaded systems (Albins 2012). Fisheries independent methods, such as visual belt transect surveys and stationary visual censuses, are appropriate (Table 5.2).

Indirect Effects on Habitat Structure and Function

The direct effects of lionfish on native fish and invertebrate communities (i.e., through predation and competition) may indirectly affect the base of marine food

Figure 5.7. Potential lionfish competitors. Lionfish share prey and habitat resources with a number of native Atlantic fish predators, such as (top, left to right) the Nassau grouper (*Epinephelus striatus*), Black grouper (*Mycteroperca bonaci*), and (bottom, left) Coney grouper (*Cephalopholis fulva*). They also inhabit similar habitats as (bottom right) Caribbean spiny lobster (*Panulirus argus*).

webs if the species consumed or outcompeted by lionfish influence the diversity and biomass of benthic communities. These “indirect effects” may result in shifts in the benthic community structure, and ultimately physical structure, of invaded habitats. Coral reefs may be particularly vulnerable to indirect effects if lionfish predation depletes populations of grazers that control coral-algal dynamics.

Indicators and protocols
The indirect effects of lionfish on habitat structure and function can be assessed by monitoring trends in the diversity and percent cover of benthic organisms in relation to native fish and lionfish density and biomass as well the physical structure of invaded habitat. Visual quadrat and photo-quadrats surveys assess benthic community composition (Table 5.3). Rugosity and depth profile measurements are used to quantify the physical structure of the marine habitat (Table 5.3).

Sampling Design and Data Management

Whether designing a new program or modifying existing ecological monitoring protocols to include lionfish and their effects, the spatial and temporal scale of sampling are critical considerations. Protocols will vary by area, depending on the characteristics of the habitat to be monitored, the goals of the lionfish control program, and the resources available. Several excellent references provide detailed information on the design and implementation of ecological monitoring methods (Tables 5.3 and 5.5) and resource managers should consult these as they develop their lionfish monitoring plans. Important considerations for monitoring design include the following:

Where to monitor
Monitoring (both for lionfish and their ecological effects) should occur in areas prioritized for lionfish control (see Chapter 4) and at similar habitats where lionfish are not actively managed. It is important to expend the effort to sample sites *not* targeted for lionfish removal, because these sites provide a reference against which the effectiveness of control/removal efforts can be compared.

Integrated sampling
Where possible, information on native fishes, lionfish, and benthic habitat structure should be collected from the same sampling unit (i.e., multiple types of data from the same survey). Integrating sampling in this way enables comparison of variables at a fine spatial scale (e.g., transects within sampling area, usually sites).

Sample size considerations
Surveys should be replicated across space and at regular intervals over time, taking into account the seasonality of organism abundance and distribution, as well as local lionfish colonization rates, and the frequency of lionfish control.

Data management
Data should be managed electronically with software such as Microsoft Excel® and Access®, or similar open source programs. Because data security is of major importance for all monitoring projects, backing up information in several physical locations and on the Internet, using an on-line data storage service (e.g., Dropbox™ or Mozy®) is key to avoiding corrupted, damaged or lost data. It is highly recom-

Table 5.3 In-water methods to monitor potential lionfish prey and competitor fish species, and benthic composition. Selection and implementation of the methods described below depend on habitat type, monitoring resource considerations, and the types of ecological effects to be monitored. Detailed descriptions of each protocol and its implementation can be found in "Ecological resources" shown in Table 5.5.

Ecological effect	Indicator	Component	Protocols	Metrics	Training required
Predation (direct effect)	Trends in diversity, density, and biomass of prey-sized invertebrates and fish in relation to changes in lionfish density and biomass	Fish prey	Visual transect surveys	Diversity, density, and biomass	Visual identification of Atlantic and Caribbean fishes to the species level Familiarity with conducting visual transect survey methodology
		Invertebrate prey	There is currently no standard visual survey method		
Competition (direct effect)	Trends in the diversity, density, and biomass of large-bodied macro-invertebrates, and predatory fish (i.e., >20 cm total length) over time in relation to lionfish density and biomass	Fisheries independent	Visual transect surveys or stationary visual surveys	Diversity, density, and biomass	Visual identification of Atlantic and Caribbean fishes to the species level Familiarity with conducting visual transect survey methodology
		Fisheries dependent	Trends in CPUE and body size of fisheries landings over time	Diversity, CPUE, catch biomass	Visual identification of Caribbean reef fishes to the species level
Habitat structure (indirect effect)	Trends in percent benthic cover and structural complexity over time on coral reefs in relation to lionfish density and biomass	Benthic composition	Visual quadrats or photo-quadrats	Diversity, percent cover	Visual identification of Caribbean and Atlantic benthic organisms
					Familiarity with conducting visual or photo quadrat assessments
		Physical structure	Rugosity and depth profiles along transects	Rugosity and depth variability	Familiarity with conducting rugosity and depth measurements

mended that managers used the USGS Non-indigenous Species Database as a repository for all lionfish sightings information.

Analyzing ecological data
Ecological monitoring data are often spatially or temporally correlated, meaning that surveys conducted at one location or habitat, or during one time period, are more similar to each other than surveys conducted at separate locations, habitats or time periods. Because each survey is not independent, analyses must reflect the "nested" nature of the data. An excellent reference for guiding ecologists through data exploration can be found in Zuur et al. 2010.

For spatially or temporally correlated data, mixed-effects models are often most appropriate (Figure 5.8). For data that cannot be transformed to achieve normality (e.g., count and percent cover data), "generalized" forms of mixed models can be used. Several excellent references provide detailed information on sample design and statistical analysis (Krebs 1999, Schiener et al. 2001, Zuur et al. 2007).

Lionfish and Ecological Monitoring Designs

Below are two examples of monitoring plans that integrate protocols for assessing local lionfish populations and their ecological effects. The spatial design of sampling and the choice of methodologies reflect location-specific habitat type and environmental conditions.

Example 1: Continuous coral reef system
Location: Puerto Rico and St. Croix, USVI
Investigators: SFU, REEF, NOAA, University of Puerto Rico–Mayagüez (UPRM)

Sampling design
A set of monitoring protocols were implemented to track the effectiveness of lionfish removal in reducing local lionfish populations and their ecological impacts on a continuous coral reef system. The system is comprised of a 20 km long fore reef,

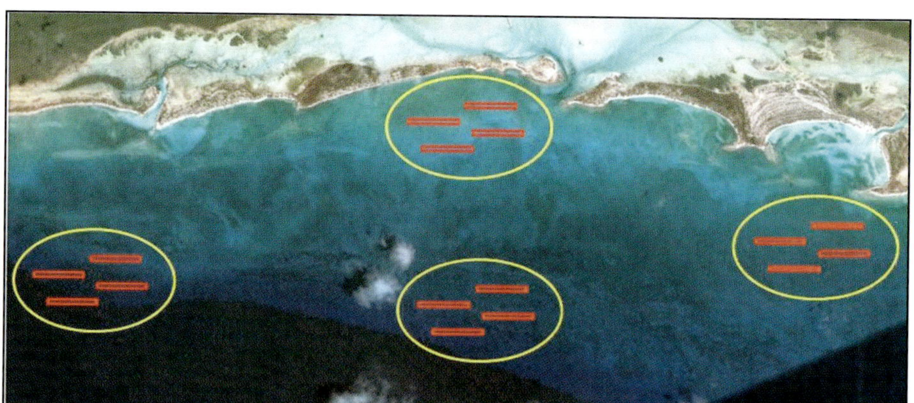

Figure 5.8 Surveys (*shown as red lines*) conducted within a location or habitat type (*the area within each yellow circle*) may be more similar to each other than those conducted a separate location or habitat type, and are therefore not independent. When conducting statistical analyses, managers must account for this "nested" data structure.

which lies along the top of a deep drop-off, ~300 m from shore. The average depth at the top of the reef crest is 9 m. The frequency of lionfish removals conducted by local divers and volunteers varies along the reef system. As a result, there may be differences over time in the establishment of lionfish and the magnitude of their ecological effect between areas which are subject to different levels of control.

Sixteen monitoring sites (each ~1 ha in area) were selected along the reef, each separated from its nearest neighbor by at least 1 km. At each site, visual surveys for lionfish, native fishes, and benthos are conducted by trained observers who were scuba diving. All data — lionfish, native fishes, and benthic habitat information — are collected along 8 belt transects of 25 m length at each site. The transects are laid parallel to the reef crest, no closer than ~5 m to the drop-off and stratified by depth and position (i.e., reef wall, reef crest, and reef flat). Monitoring at each site occurs on a biannual basis (once in winter, once in summer).

Native fish survey

While rolling out each transect line, the diver records the identity and size (total length the nearest 1 cm) of all fishes greater than 15 cm within a distance of 2 m from either side of the transect (transect area: 25 m x 4 m). The minimum time for this survey is 5 min, and increases in higher complexity sites. After waiting 3 min to reduce observer effects on fish behavior, the diver makes a second pass to record the identity, size (total length the nearest 1 cm) and abundance of all fishes less than 15 cm within a distance of 1 m from either side of the transect line. (See "*Suggested standard visual survey procedure for fish prey,*" Page 60).

Lionfish survey

After conducting native fish surveys on each 25 m transect line, the observer makes detailed searches only for lionfish within 5 m of either side of the same transects (each transect area: 25 m x 10 m), using the methods described under "*Suggested standard survey procedure for lionfish*" and in Figure 5.5. The search encompasses the same area in which native fishes are surveyed on each transect.

Benthic habitat

Data on benthic composition, rugosity, and structural relief are collected from multiple points along each 25 m transect line. Photo quadrats are conducted every 3 m (for a total of 8 photos per transect), and the camera distance from the reef is standardized (e.g., 50 cm) to ensure a consistent area for analysis. Reef rugosity is measured every 5 m along the transect line (for a total of 5 measures per transect). To assess rugosity, divers fit a fine-linked 3 m long chain to the substrate perpendicular to the transect line, paying careful attention to fit the chain into all contours of the benthos. The straight-line horizontal distance between the transect tape and the end of the 3 m chain is recorded to create a ratio of the fitted distance to straight-line distance (i.e., rugosity). At the same points (every 5 m), the height of the reef structure (i.e., relief) is recorded by measuring the depth of the lowest and highest point within a 1 m radius of the transect tape. The depth of the reef structure directly under the transect tape at each 5 m point is also recorded.

Example 2: Coral patch reef system
Location: Eleuthera Island, Bahamas
Investigators: Cape Eleuthera Institute, SFU, REEF, Bahamas Department of Marine Resources

Sampling design
A set of monitoring protocols were implemented to track the effectiveness of lionfish removal in reducing local lionfish populations and their ecological impacts on coral patch reefs within a large, shallow sound. The system is comprised of numerous patch reefs varying between 100–150 m^2 in area and separated by between 200 m –1 km of sand and seagrass habitat from its nearest neighbor. The depth of the water in the sound is ~4 m. The frequency of lionfish removal varies between patches, so that there may be differences in lionfish colonization (i.e., density and biomass) and their ecological effects between sites with higher and lower control. In this system, each patch is considered a monitoring site, and a subset of 32 patches, varying in removal frequency, are selected for monitoring. At each site (i.e., patch), visual surveys for lionfish, native fishes, and benthos are conducted by trained observers diving with scuba gear.

Because the patches represent small, discrete habitat units, two types of surveys are used for monitoring. Lionfish, native predatory fishes, and macro-invertebrates are censused during roving diver surveys of the entire patch. Native fishes and benthic habitat are characterized on 4 belt-transect surveys at each patch. Transects are oriented north-south on the patches. Three belt transects are also laid in the sand/seagrass adjacent to each reef (~5–10 m from the reef edge), to assess fish species in these adjacent habitats. Monitoring at each site occurs on a biannual basis (once in winter and once in summer).

Whole-reef lionfish and predatory native fish surveys
Prior to placing transect lines, the surveyor makes a systematic search of the entire patch reef area, recording the size (total length to the nearest 1 cm) of all large predatory fish species (Serranidae and Lutjanidae). For all lionfish located on the patch reef, the surveyor records total length (to the nearest 1 cm). Search time will vary with patch size, but should be no less than ~15 m^2 per minute.

Transect surveys for native fishes
A diver records the identity and size (total length the nearest 1 cm) of all fishes within a distance of 1 m from either side of the transect (transect area: 8 m x 2 m). The time for each survey should be standardized to ~8 minutes, but may be longer for highly complex habitats (modified from Page 60, "*Suggested standard visual survey procedure for fish prey*"). A diver also records the identity and size of fishes along the 3 seagrass/sand transects, standardizing survey time for to a minimum of 3 minutes each.

Benthic habitat surveys
Data on benthic habitat composition, rugosity, and structural relief are collected from multiple points along each 8 m x 2 m transect, using the same methods as described above. Photoquadrats are conducted every 1 m (for a total of 8 photos per transect). Rugosity, relief, and depth are measured every 2 m along each transect line (a total of four measures per transect).

ORGANISMAL MONITORING THROUGH DISSECTION

Data generated through lionfish dissection can inform our understanding of their biology, ecology, and potential impacts on invaded habitats. Coupling dissection information with environmental and biotic data facilitates inferences about lionfish population dynamics, reproductive biology, and food web ecology across space and over time.

This section briefly describes the dissection data required to inform the management of lionfish and the impacts on invaded marine habitats. Ideally, data and samples from specimens would be collected in a standardized way to facilitate inter-region comparisons over time.

Standard techniques for collecting information and samples from dissected lionfish are summarized in Green et al. (2012b). Table 5.4 outlines several areas in which data obtained through dissection can be used to inform our understanding of the population dynamics of lionfish, their potential ecological effects on invaded marine habitats, and the effectiveness of control.

Population structure
The structure of populations can be determined through genetic approaches, which use tissue samples collected from dorsal fin, caudal fin, gill, and muscle tissues.

Population dynamics
Age and growth information from individual fish can be determined using otoliths (ear stones) in conjunction with fish body lengths. The relationship between weight and length of lionfish specimens can also be used to monitor changes in body condition, which may correlate with environmental variables such as: habitat type,

Table 5.4 Types and uses of data collected through lionfish dissection.

Type		Metric	Units	Application
External	Measurement	Total length (TL)	mm	Growth, body condition, population size structure
		Standard length (SL)	mm	
		Weight	g	
		Gape width	mm	Feeding ecology
		Gape height	mm	
	Sample	Gill tissue	—	Species identification, population structure
		Muscle tissue	—	Food web ecology through stable isotope analysis, species identification, population structure
		Fin clip		Species identification, population structure
Internal	Measurement	Gender	M or F	Individual gender, gender ratio in population
		Interstitial fat volume	ml	Health assessment
		Stomach contents	mm & ml	Feeding ecology
	Sample	Otoliths	—	Age and growth
		Gonad	—	Reproductive physiology
		Stomach contents	—	Feeding ecology

resource availability, changing latitude, or invasion status. These data can also be used to determine temporal and spatial changes in age structure of local lionfish populations.

Reproductive biology
The reproductive status of an individual lionfish can be determined using various metrics such as gonad weight, morphology, and histology. These data aggregated over space and time can yield population-level trends in reproductive biology, such as spawning frequency and seasonality, and maturation schedules.

Food web ecology
Lionfish diet can be characterized by identifying and measuring the stomach contents of collected specimens. Conducting stomach contents analysis across temporal and spatial scales can improve our understanding of changes in diet composition and potential community effects. The carbon and nitrogen isotopic signature of lionfish muscle and fin tissues can be used to assess trophic position and feeding ecology. These data can be used in conjunction with isotope 2 data from native species to describe biotic interactions between lionfish and members of the invaded food web.

MONITORING SOCIOECONOMIC IMPACTS: FISHING, TOURISM, AND HUMAN HEALTH

The economies and societies of many countries across the western Atlantic, Caribbean, and Gulf of Mexico depend greatly on fisheries and tourism industries. These two major sectors may be affected by the ecological impacts of lionfish. A third concern is the health risk to people who come in direct contact with venomous lionfish. This section offers resource managers a general framework for assessing the economic, social, and human health effects related to lionfish. To date, few efforts have been made to quantify the magnitude and prevalence of socioeconomic impacts from the invasion. This section will be updated as new information and techniques become available.

Sampling design
Monitoring the socioeconomic effects of invasive lionfish requires linking social and economic data to the prevalence (i.e., abundance, density, and biomass) of lionfish within the area, both across space and over time. Information on the status of local lionfish populations can be obtained through both fisheries dependent and independent sampling methods (See Ecological Monitoring). Managers should relate these data to the social and economic metrics described below.

Fisheries

Lionfish may affect the catch per unit effort of fished species in two ways: 1) they may directly affect the population size of the target species (through predation and/or competition), and 2) they may interfere with fisheries activities, so that the cost of fishing increases – in terms of time, effort, or safety risk (Figure 5.9). These effects will likely vary by fishing method (gear type), target species, habitat, and the density of lionfish in the area.

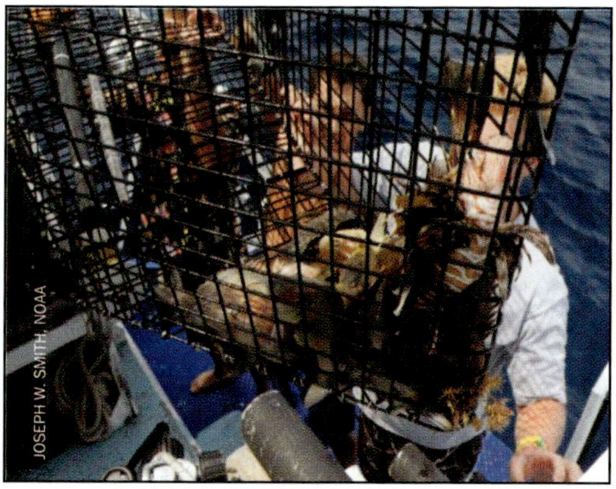

Figure 5.9. Lionfish caught in lobster and fish trap fisheries around the Caribbean raise concerns about their effects on the catch of commercially valuable species and the risk of envenomation to fishers. In particular, there is concern that the extra time and effort required to handle venomous lionfish caught within traps is reducing fishers' efficiency in collecting their catch (i.e., reducing the catch per unit effort).

Indicators

i) Catch-per-unit-effort (CPUE) and total catch of target species, and
ii) Bycatch-per-unit-effort and total bycatch over time of lionfish.

Protocols

i) Fisheries dependent — observation of commercial and subsistence fishing activities; minimum data to be collected:
- Location, habitat type, and depth of fishing
- Number of fishers, gear type, and duration of fishing
- Expenses per day of fishing (e.g., fuel, supplies, wages)
- Total number, size, and identity of target species caught
- Total number, size, and identity of bycatch (including lionfish)

ii) Fisheries independent — research fishery conducted by resource management staff:
- Semi-structured interviews of fishers' perception of lionfish effect on catch and livelihood
- Same data to be collected as above

Tourism

While quantifying the economic effects of lionfish on the tourism industry will take long-term effort, gathering data on the perceptions of visitors and vendors is a valuable start to assessing impacts. There are two main ways in which the invasion could affect tourism:

i) Change to the structure and function of ecosystems regularly visited by tourists. Ecosystem change could affect tourism negatively if these systems are no longer desirable attractions. For example, a reduction in the diversity and density of fishes on coral reefs or reduced coral cover may affect dive tourism; reduced populations of recreationally fished species, such as bonefish, may affect international fishing tourism. However, the presence of lionfish colonization could boost tourism if sighting the species (i.e., during snorkeling or scuba diving) or fishing for them are desirable activities.

ii) Increased risk to human health for tourists accessing invaded diving/snorkeling sites or near-shore beach areas because of the possibility of envenomation by lionfish.

Recreational Diver Perception of Lionfish Impacts

Dive masters, diving instructors, and tour operators are a vast source of qualitative information on changes in reef health and lionfish populations, as well as on tourist perception of the invasion and its impacts. Interviews with these dive professionals can provide anecdotal information on whether the lionfish invasion is affecting dive-tourist experiences and the likelihood of divers returning to a lionfish invaded location. Similar surveys have been conducted to assess the impacts of coral bleaching on dive tourism in the region (Cesar et al. 2000, Sealey-Baker 2011).

A survey targeting the dive-tourist experience in lionfish -invaded regions would focus on three key questions:

i) Do dive/snorkel tourists notice the lionfish,
ii) Do the lionfish adversely impact the enjoyment of their dive, and
iii) Does the presence of lionfish affect the number/location of dives they are making or their intent to return to the same area?

Indicators

i) Rate of envenomation among tourists,
ii) Diver perception of the effects of lionfish on coral reefs,
iii) Changes to rates of capture of recreationally valuable species, and
iv) Tourism revenue over time.

Protocols

i) Envenomation records from medical sector,
ii) Semi-structured interviews with the tourism sector and tourists,
iii) Surveys (online), and
iv) Economic data collection (tourism revenues audit).

Human Health

Lionfish are venomous and there is a risk that people who encounter the fish will be stung. Sectors of society who are in close proximity to lionfish will likely be at a greater risk. These include fishers, divers, beach-goers, and those in the restaurant trade where lionfish are used as a food fish. As the density of lionfish in an invaded area increases, there is a greater risk that people from these sectors will encounter lionfish, and potentially an increase in the risk of being stung.

Indicators

i) Rate of envenomation among different societal groups,
ii) Cost of treatment to health system, and
iii) Trends in long-term effects on individuals who have been stung.

Protocols

Envenomation records and reports from health officials, including information on:

i) Location, date and time of envenomation,

ii) Activity at time of envenomation,
iii) Demographic information,
iv) Symptoms,
v) Follow up for long-term effects, and
vi) Cost of treatment.

Example
As recommended within their National Lionfish Response Plan, the Commonwealth of the Bahamas has developed a questionnaire aimed at tracking the prevalence, severity, and distribution of lionfish envenomation across the country (Appendix 5). The questionnaire is intended for physicians at both hospitals and clinics. The questionnaire is still pending approval by the Department of Health and as a result is not in use at this time.

Information Resources

Whether designing a new program or modifying existing ecological or socioeconomic monitoring protocols to include lionfish and their effects, resource managers should consider as critical the spatial and temporal scale of sampling. These details will vary by area, depending on the goals of the control program and resources available in each region, and are thus beyond the scope of this section. Several excellent references provide detailed information on the design and implementation of ecological and socioeconomic monitoring methods (Table 5.5) and managers may consult them as they develop their own plans for lionfish monitoring.

Table 5.5 Sources that provide detail on the design and implementation of monitoring methods.

Resource type	Reference	Open access link
Ecological	Hill Wilkinson (2004)	http://www.icran.org/pdf/Methods_Ecological_Monitoring.pdf
Ecological	Menza et al. (2006)	http://www.ccma.nos.noaa.gov/publications/Reefmonitoringguide.pdf
Ecological	Rogers et al. (1994)	http://fl.biology.usgs.gov/Monitoring_Manual.pdf
Ecological	Labross et al. (2002)	http://www.spc.int/DigitalLibrary/Doc/FAME/Manuals/Labrosse_02_UVC.pdf
Ecological	Caldow et al. (2009)	http://ccma.nos.noaa.gov/products/biogeography/fgb/
Ecological/ Socio-economic	Wilkinson et al. (2003)	http://www.reefresilience.org/pdf/mcrmpa-v1.pdf
Ecological/ Socio-economic	Sullivan Sealy et al. (2006)	http://henge.bio.miami.edu/coastalecology/TOOLS%20&%20METHODS%20April.pdf
Socio-economic	Bunce et al. (2000)	http://www.socmon.org/pdf/GCRMN_Manual.pdf
Socio-economic	Bunce and Pomeroy (2003)	http://www.socmon.org/publications.aspx

CHAPTER 6

LEGAL AND REGULATORY CONSIDERATIONS FOR LIONFISH MANAGEMENT

Dayne St. A. Buddo

The lionfish invasion has spawned an array of legal challenges. Lionfish management in every jurisdiction can encounter legal and regulatory issues, which must be addressed to ensure the success of the management program. In some instances, existing laws and regulations may assist in the management of this invasive species. However, the more important issues are the legal and regulatory instruments that counteract management strategies for control of the species. One such issue is the removal of fish from marine protected areas (MPAs) that specifically prohibit fishing. Depending on the legal system in a jurisdiction or specific law, resource managers need to include considerations for regulatory amendments or even new legislation in their plans, if they intend to engage in and/or promote lionfish removal activities.

The intent of this chapter is to provide guidance to resource managers and policymakers on the legal and regulatory challenges surrounding the lionfish invasion. This guidance will assist in the preparation of response plans and legislative instruments for effective control throughout the Wider Caribbean and other regions affected by the invasion. It is expected that the policies and perspectives provided here would need to be adapted, in time, to address the specifics of each jurisdiction. Inevitably, multiple stakeholders should work to reach consensus and implement strategies and practices in the context of a coordinated regional approach.

To date, no single legislation comprehensively addresses the lionfish invasion. Some countries have chosen to develop special provisions within existing legislation, mostly as are related to the removal of the lionfish in regulated areas.

CONSIDERATIONS FOR LEGAL AND REGULATORY INSTRUMENTS

This section offers guidance for addressing legal and regulatory aspects of lionfish management at various levels of stewardship, from local marine park managers to those at the regional scale. These considerations were developed from the insights of partners and stakeholders in the Wider Caribbean and can be used to help develop local policy statements, legislative drafting, or even other management actions and practices. These are the main considerations as voiced by various countries throughout the region.

Consideration 1: Removal of lionfish from no-fishing areas

There has been a regional trend over the last decade to designate marine areas as *no-take* areas, primarily so they can serve as fish nursery grounds. In these areas, fishing and, by extension, the removal of any marine life are prohibited. These areas often show high abundances of juvenile fish and crustaceans, as is typical of breeding/spawning areas. However, this causes a regulatory conflict. Since no fishing or removal is allowed, lionfish removal from these ecologically sensitive areas also is not allowed. This effectively provides a safe haven for lionfish.

The removal of lionfish from *no-fishing* areas – such as fish sanctuaries, marine parks, and other protected areas – could be considered a specially allowed activity. Exceptions to existing laws and regulations governing gear type and the activity of fishing could be explored in order to review deviations from the existing regulations.

Special permits under the respective laws could be issued, including guidelines and sanctions that:

i) Refer to removal of lionfish only,
ii) Supervise persons conducting removal (by government or officers),
iii) Specify dates and times for removals,
iv) Prescribe methods and gear types for removal,
v) Address safety considerations,
vi) Propose sanctions and/or fines for the violation of the guidelines,
vii) Notify local fishers that lionfish removal would take place at a particular date and time, and
viii) Include provisions for reporting and final disposition of specimens.

Special consideration should be given to granting these permits to local fishers, to avoid potential conflict and to reduce the need for enforcement.

Bermuda and Jamaica have implemented legislation and regulations that provide a mechanism to allow removal of lionfish in otherwise no-fishing areas. In Jamaica, the previously declared fish sanctuaries in 2009 have been re-designated in 2012 as Special Fishery Conservation Areas. This now enables officials to grant special permits, allowing the removal of lionfish from these areas. The Cayman Islands have also developed regulatory instruments for lionfish control; these are detailed later in this chapter.

Consideration 2: Importation and/or use of lionfish for the aquarium trade
The broad consensus is that lionfish were introduced in the Atlantic from the aquarium trade. It is uncertain if this introduction was accidental or purposeful. The continued importation (regionally or nationally) and trade of live lionfish is a subject that demands careful consideration and debate. Some locations are considering banning possession or trade of lionfish (*Pterois volitans* or *P. miles*) for marine aquaria. Many countries of the Wider Caribbean region have policies restricting trade or import of live specimens of non-native species. Some locations are permitting live lionfish displays for research and public education activities only.

Careful consideration should be given to the use of the lionfish in the pet/aquarium trade. Such considerations should include the destination of the live lionfish, special requirements for the disposal of unwanted lionfish to prevent human-mediated spread, the potential benefits to the environment through increased removal of juveniles for the aquarium trade, the socioeconomic benefits and dependence created for income by local fishers, and the possible negative public perception of encouraging/promoting an activity that caused the lionfish invasion in the first place.

There are a myriad of ways of looking at the issue. In the Bahamas, for example, placing a ban on lionfish for the aquarium trade is of low priority under the Bahamas National Lionfish Response Plan (Appendix 6). This relates to the use of lionfish taken from Bahamian waters. This was viewed as an activity that would require significant resources to enforce, while the benefits may be negligible.

Consideration 3: Use of lionfish as a fishery resource

There is no broad consensus regarding the advantages and risks of promoting a commercial market for lionfish. On one hand, lionfish flesh is widely recognized as excellent, and economic forces (i.e., the connection of supply chains with potential market demand) may be powerful tools in controlling lionfish populations in high-priority locations. On the other hand, some risks accompany this approach, including potential secondary impacts of increased fishing pressure, unknown market requirements for sustainable trade, impacts of new fishing practices on fragile marine ecosystems, challenges in regulating increased fishing effort, and the possibility of creating perverse incentives to introduce or maintain lionfish populations as an economic resource.

While additional exploration and dialogue is clearly necessary, several possible approaches to a lionfish fishery include:

i) Marketing the lionfish as an "eco-friendly" fishery resource with the intent of reducing the impact on the other local traditional fisheries,
ii) Special promotion of an "Appetizer" version (i.e., the smaller and sexually immature lionfish) in addition to the larger, sexually mature "Entrée" version,
iii) Implementing a special permit to commercial entities to use lionfish for commercial sale, and
iv) Use of licensing fees for, or percentage of sales going directly to, the management and control of lionfish. This message can also be clearly outlined on the packaging of the processed lionfish and/or be featured in advertisements and other marketing activities.

Lionfish bioaccumulation of ciguatoxins has been a concern for many countries in the region that promote consumption as a control strategy. This has been of greatest concern in countries described as "hotspots" for ciguatera poisoning.

Seafood Company Supports Lionfish Project

In March 2011, in Jamaica, one of the largest seafood processing companies in the region, Rainforest Seafoods Co. Ltd., partnered with the National Lionfish Project. This company is very concerned about the impact that the lionfish is likely to have regionally, especially in the countries from where they get their supplies of seafood. The company has acknowledged that one of the ways to help control the lionfish is to increase its consumption by humans. They have embarked on an effort to commercialize the use of lionfish, primarily from the standpoint of conserving the fish being consumed by the lionfish, thereby protecting their already established market for these prey items. They are exploring the use of the juvenile lionfish as well. This effort aims to exploit lionfish as a commercial fishery resource in a very unsustainable manner to reduce its numbers.

Consideration 4: National and regional coordination
National coordination on lionfish management provides an essential link between broader regional coordination and the personnel in the field in that country. It is useful to adopt best practices consistent with those being implemented throughout the region, and equally important for national adaptation of best practices suited to their respective needs. Some activities for consideration may include establishing a national lead/coordinator and a national committee or task force.

Lionfish do not recognize boundaries of countries, jurisdictions, or marine management areas, and therefore, it is essential that management and policy responses are synergized regionally, at the ecological scale of the invasion. Policies will be more effective when they are consistent, or at least complementary, across boundaries. Similarly, monitoring and research activities will be more powerful if fundamental methodologies and data collection standards are shared. There is an immense wealth of experience across the region that should be shared, so that individual managers and policymakers do not have to re-learn the same lessons in different places.

The managers who assembled in Cancun in August 2010 emphasized these principles, and many efforts are now underway to increase communication, collaboration, and coordination in response to the lionfish invasion.

REVIEW OF EXISTING LEGISLATION, POLICIES, AND PLANS

The following sub-sections provide information on some of the legal and regulatory measures taken by countries throughout the region.

Puerto Rico
In November 2010, Puerto Rico created a "Special Disposition for the Lionfish" under Fish Regulation 7949 (Appendix 7). This special provision provides allowances by permit for the live and dead capture of lionfish in any marine environment, including no-fishing/no-take and other protected areas. It also makes provisions for use of scuba, hookah, or any type of gear to catch lionfish during day or night. In addition, the Puerto Rico Department of Natural and Environmental Resources established a special program to administer the special allowances. This important component provides organizational structure ensuring a level of regulatory control on the removal of lionfish.

The Bahamas
In 2003, the Bahamas developed a National Invasive Species Strategy (NISS) addressing the threat of terrestrial and aquatic invaders. Given that the lionfish invasion was not documented in the Bahamas until 2004, no special provisions for lionfish were addressed in the NISS. In 2009, based on recommendations from the Bahamas NISS, a National Lionfish Response Plan (NLRP) was developed with leadership of the Bahamas Department of Marine Resources (Appendix 6). Recommendations from these documents have led to the proposed amendment of several regulations in Chapter 244 of the Fisheries Resources (Conservation and Jurisdiction) Regulations. These recommendations include amending existing legislation or creating new legislation that addresses the management of invasive species, creating and updating priority lists of species for eradication, and suggesting that

the Government of the Bahamas adopt an Invasive Species Policy. These recommendations are currently under review.

One of the main objectives of the NLRP is to develop appropriate policies and regulations via collaborative efforts among local and regional public and private sector partners. Three tiers of action for managing the lionfish invasion are provided, including amendment of legislation to allow lionfish removal, encouragement of the taking of unlimited numbers of lionfish for consumption, and the expansion and protection of the system of Marine Protected Areas (MPAs). The lowest priority (with an action date within ten years) is to ban lionfish as aquarium pets and restrict possession or transport of live lionfish. Recommendations specific to lionfish harvesting encompass special permission to hunt lionfish in no-spearfishing areas, the use of a variety of gear (including scuba and hookah), the holding of lionfish tournaments, and the allowance of visitors to take unlimited quantities of lionfish out of the Bahamas. These amendments were developed to maximize lionfish removal in the shortest amount of time. Note that the Government of the Bahamas has yet to implement these amendments.

The Cayman Islands

The Cayman Islands developed three separate, but related, regulatory instruments for lionfish control to be implemented by the Cayman Islands Department of Environment (Appendix 8). While these are not formal pieces of legislation, the Cayman Islands Department of Environment (DOE) has implemented these guidelines with support from the Marine Conservation Board (MCB), the authorized body to make exceptions to the Marine Conservation Law. These exceptions were executed under special circumstances, of which a lionfish invasion qualifies.

In 2009, the MCB approved and began issuing a special license to permit removal of lionfish from two of three MPA types, including marine park zones and replenishment zones (but not the environmental zone). Prior to the lionfish invasion, the Cayman Islands had prohibited any use of spears, and as such, only nets were allowed under this license. Exceptions were made to the fishing regulations, permitting:

- i) Capture of lionfish using gloves and scuba in taking marine life,
- ii) Taking marine life from the marine parks and marine replenishment zones, and
- iii) Taking fish less than eight inches in fork length.

A reporting system is also a feature of this system.

In 2010, the Cayman Islands Tourism Association expressed an interest in using dive staff at hotels to remove lionfish. The MCB granted this permission, and through the DOE, established a dive company spearing program (Appendix 8). Given that it was illegal to import spears and components of spears into the Cayman Islands, manufacturing of pole spears in the Cayman Island became necessary. This program provided guidelines for the use of the spears exclusively issued by the DOE and only by the licensee. Return of the spear to the DOE is required after use. If licensee does not adhere to conditions of the license, spearing privileges may be revoked and prosecution may follow.

In March 2011, a legal change was executed at the request from residents to begin a resident culling program in the Cayman Islands. In addition, there was greater need for more effective control strategies given the number of lionfish. Special permission was granted by the government to allow the importation of short polespears from the United States. This program has since been implemented and includes mandatory participation in a training program on lionfish removal by the DOE (Appendix 8).

Mexico

The Parque Nacional Arrecifes de Cozumel implemented a Volunteer Lionfish Capture and Control Program promoted by the Comisión Nacional de Áreas Naturales Protegidas (CONANP). This structured program allows interested persons to participate in the removal of lionfish using scuba. CONANP implemented a waiver of liabilities (Appendix 9) acknowledging the risks associated with diving on scuba to hunt for lionfish.

French West Indies

In the French West Indies, an invasive alien species (IAS) is placed on a nuisance species list when it is deemed to negatively affect agriculture and farming. Species on this list are then the subject of actions designed to remove it from the wild or farmed areas. France does not presently consider lionfish an IAS; hence, there are no regulations regarding the invasive species. Lionfish are subject to the Environment Code, however, which allows the administration to capture, take, keep or destroy the specimens of introduced species when their presence in the wild is documented. All IAS species are covered by this provision, but its enforcement is restricted to nominative lists of species decided by the Minister of Environment (Appendix 10). In early 2011, a local decree was passed in Guadeloupe and Martinique authorizing the capture of lionfish by scuba divers with a gun or special kit. Currently, there are only approximately 50 persons with this authorization throughout the French West Indies.

U.S. Marine Protected Areas

In the United States, the National Park Service (Department of the Interior) and the National Marine Sanctuaries Program (NOAA) have developed response plans to mitigate the impacts of lionfish. These include encouraging the harvesting of lionfish as a food fish, development of park- and sanctuary-specific lionfish control plans, and the review of regulations that may inhibit lionfish removals. In the Florida Keys National Marine Sanctuary, permits are being issued to allow the use of prohibited gear in the sanctuary. Permit holders are required to attend a special training aimed at reducing the impacts of removal activities on the reef and protecting human health.

Bonaire

The Bonaire National Marine Park has special prohibitions for removal of any marine life within the marine park (Appendix 11). This includes commercial, artisanal, and recreational fishing. It also speaks to prohibition of gear type, where even the possession of that gear or parts of the gear is an offense. However, the law provides for exceptions to these aspects for the manager and agents of the park. Therefore, under existing laws, the marine park management was able to allow the removal of lionfish.

Chapter 7

RESOURCES, PARTNERSHIPS, AND SUSTAINABLE FUNDING

Ricardo Gómez Lozano

The development of a comprehensive strategy for securing resources and utilizing partnerships is critical to a successful management approach to the lionfish invasion. Some activities for lionfish control can be incorporated into existing program budgets and work plans. Most activities, however, will require new and renewable funding sources in order to be effective. Sources for sustainable funding include establishment or earmarking of special funds for lionfish management, contributions from beneficiaries, and development of a trust or mitigation-type fund.

One of the key elements in securing funding is a sound management plan with prioritized actions. These actions may be prioritized based on the potential effectiveness of each one in mitigating lionfish impacts. A successful program also has several components, including planning, motivation, knowledge, funding, staff capacity, and action. The absence of any of these components can interfere with constructive outcomes. In many locations, government agencies have taken the lead role in invasion control and management, with strong support from non-governmental and other research entities.

This chapter presents a number of issues that should be addressed in a resource acquisition and allocation strategy. This list is not intended to be exhaustive but to provide a beginning to guide managers in the planning process. Before initiating a lionfish research and control program, resource managers must have a clear idea of the resources necessary to start the program and sustain it for the long term.

HUMAN RESOURCES

Consider that, in most Caribbean countries, lionfish have invaded within the last five years. Without question, the lionfish invasion has caused a major shift in staff allocation to marine resource management in the Caribbean with individual staff time devoted to this problem ranging from 10–85% in some locations (Table 7.1). A useful way for resource managers to address and meet the demand of human resources is to clearly identify existing and new skills sets and evaluate organizational needs. In each institution, a manager can generally find staff members with the skills and aptitudes needed to carry out new or different roles.

Roles and responsibilities

The first rule in planning an effective response effort is to get organized! The organization(s) taking the lead must decide who participates in the control actions, when is the best time to execute the actions, where is the best place to perform the activities, and how to involve different stakeholders and decision makers in lionfish management. Below is a list of potential roles of personnel within a lionfish program. It is intended to assist managers in compiling human resources to address the lionfish invasion. The proposed roles can be adapted to specific locales, and one person may have to assume multiple roles, depending on local staffing resources.

i) *General Coordinator* — This person is responsible for managing all activities arising in the actions of outreach, education, control, and research, as well as the economic resources necessary for their development. An important part of the general coordinator's responsibility is to know the legal framework, the authorities, and duties, before developing the lionfish control program. The coordinator has the most direct contact with staff and is responsible for field activities and coordinating with partners and volunteers. The coordinator also defines roles and responsibilities and establishes an organizational chart.

ii) *Outreach and Education Coordinator* — An effective control strategy should include a public outreach and education program to raise awareness about invasive species and their impact on native resources. The person responsible for this program would oversee the preparation of printed and electronic materials, determine appropriate means for dissemination, and establish messaging that promotes control and possible use and exploitation of lionfish. The outreach and education officer must also have the skills to train staff and the subject-matter expertise to develop factual outreach materials.

iii) *Research and Monitoring Coordinator* — An effective control strategy should consider monitoring and research to better understand the status and behavior of local lionfish populations. Establishing baselines and conducting periodic assessments are necessary to determine lionfish population size and growth rate, its impact on native species, and the effectiveness of control strategies. To do this, technically and scientifically trained personnel are needed to coordinate efforts with existing staff and collaborators within the local government, universities, and other research institutions.

Human resources from the local community

All community members that manage or utilize coastal and marine resources are capable of contributing to the control of lionfish. For example, fishers can develop strategies that enable them to reduce the impacts of lionfish in fisheries by targeting lionfish or retaining bycatch, and governments can develop strategies to mitigate impacts on tourism and fisheries through development of fishery management plans. Tourism industries can be involved directly in the control efforts through adopt-a-reef programs. Basically, all community members that generate economic gains derived from marine resources can contribute resources to support control programs. Thus, it is important to identify local human and economic resources to best achieve program goals.

Table 7.1 Estimates of the impacts of lionfish on coastal manager staff time in the wider Caribbean region obtained through a questionnaire on the GCFI listserv (2011).

Organization	Country	First lionfish sighting	# of staff	% staff time on lionfish	Estimated staff working on lionfish
Belize Fisheries Department	Belize	2008	72	10%	7
Trinidad & Tobago, Institute of Marine Affairs	Trinidad & Tobago	Unknown	100	4%	4
Cayman Islands Department of Environment	Cayman Islands	2008	33	8-10%	3
Puerto Rico Department of Natural and Environmental Resources	U.S., Puerto Rico	2008	1500	10%	150
St. Lucia Department of Fisheries	St. Lucia	Unknown	30	At least 50%	15
STINAPA Bonaire–Bonaire National Marine Park	Bonaire	2009	25	<5%	1
Bahamas Department of Marine Resources	Bahamas	2004	55	10%	6
Acuario Nacional de Cuba	Cuba	2007	—	40%	
Old Providence Ecohamlet Foundation	Colombia	2008	6	—	
Reef Check República Dominicana	Dominican Republic	2008	3	5%	1
Ministerio de Medio Ambiente y Recursos Naturales	Dominican Republic	2000	8	10%	1
University of the West Indies – Discovery Bay Marine Laboratory	Jamaica	2008	20	>50%	3
National Environment and Planning Agency	Jamaica	2008	400	>50%	10

Capacity building
It is important to consider that an effective control and research strategy must be implemented on a regular basis and over the long term. For this reason, it is necessary to continuously train personnel, as they respond to different stages of the invasion. A *train-the-trainers* approach, where staff members are trained to train others, is an effective method for ensuring capacity building as well as the continuity of operations.

As the lionfish invasion progresses, management strategies will evolve, which will include resource managers giving consideration to new partners and new methods of lionfish control. In this regard, the following items are examples of management issues that should be periodically reviewed and updated to better achieve the strategy goals.

FINANCIAL RESOURCES

Both financial and human resources will likely be limited, thus the control strategy goals and activities should be clearly defined and prioritized. Concrete actions, specific deadlines, and evaluation are a few of the components that are essential to creating a program that will have a real impact.

Resource allocation
Allocation of resources for each program component is perhaps one of the most difficult decisions a manager will face. Over-allocation of resources in one area can compromise the effectiveness of the overall program. Managers are required to balance allocation of resources with need, which will change often – requiring nearly constant evaluation of all aspects of the lionfish program.

Below are a number of organizations that work on invasive species in the Wider Caribbean region. Some of them have information that can help in developing funding proposals and/or they report on possible funding sources; and some organizations have directly supported actions to lionfish control.

i) Convention on International Trade in Endangered Species of Wild Fauna and Flora (CITES); www.cites.org,
ii) Invasive Species Specialist Group (ISSG); www.issg.org,
iii) Centre for Agricultural Bioscience International (CABI); www.cabi.org,
iv) Specially Protected Areas for Wildlife (CAR-SPAW-RAC); www.car-spaw-rac.org,
v) International Coral Reef Initiative (ICRI); www.icriforum.org,
vi) Coral Reef Alliance (CORAL); www.coral.org,
vii) Reef Environmental Education Foundation (REEF); www.reef.org,
viii) U.S. Geological Survey (USGS); www.usgs.gov,
ix) National Oceanic and Atmospheric Administration (NOAA); www.noaa.gov, and
x) The Nature Conservancy; www.nature.org/initiatives/invasivespecies.

Resource acquisition — designing a strategy for financing

Universally, the amount of funding that can be acquired to support conservation projects is small relative to need. Programs that do not have a focused strategy run the risk of distributing the resources in an inefficient way. Partners are essential to strengthen the organization and to provide matching support. General financing must be directed to selected activities to achieve a strategic impact.

Some important points to consider are:

i) What is the nature of the environmental threat?
ii) What is the nature and duration of the conservation activity?
iii) Which other organizations are dealing with this threat?
iv) What are the strengths and weaknesses of partner organizations?
v) Is there need to establish a mechanism for governmental and non-governmental organizations to work together on this issue?
vi) How committed are the government and other key organizations?
vii) What practices are legal in the country?
viii) What practices are reliable and inspire confidence in institutions at national and international levels?

There are many resources to guide fundraising efforts. For example, Bath (2011) provides an excellent guide to identifying and managing funds for environmental capacity building (http://toolkit.conservationfinance.org/sites/default/files/documents/redlac-capacity-building/redlacfundraising-strategies-environmental-funds08112011.pdf).

To submit funding proposals to national or international organizations, it is important to consider the following aspects, which are required in most applications resources.

i) Commitment and availability of all key groups involved (e.g., council, governments, and the community),
ii) A clear vision and a strategic plan focused on the program's improvement and growth,
iii) Objectives based on clear priorities and plans, budgets, and precise needs,
iv) A clear cause to fund, properly documented and supported, and
v) A market survey of potential national and international donors, whose priorities for the granting of subsidies correspond to the defined needs or objectives of the strategy.

Types of financing

Capital funding for control efforts may come from a variety of sources. For example, the creation of *trusts* is an act of faith and confidence that allows the flow of funds between the donor and grantee. In several countries, these trusts are instruments of widespread use because government institutions may not be eligible to receive direct economic resources.

Donations are another way of delivering to and receiving economic resources and material goods by institutions of social welfare and preservation of the environment. These may range from small donations from individuals to large donations from companies concerned about the lionfish invasion.

Sponsorships are also a popular method for private entities to gain publicity through their support for environmental and conservation initiatives. Both the control program and the company's product or brand may benefit from this type of mutually supportive relationship.

Another way to obtain funding to implement control and management programs is through the *sale of materials* that refer to the invasive species, including T-shirts, hats, and handcrafts made from by-products of the organism.

Below is a summary of potential sources of funding:

i) Donations from individuals in the country,
ii) Donations from national or international companies operating in the country,
iii) Fines collected from environmental damage settlements,
iv) Shares from nature conservation taxes collected at ports of departure,
v) Domestic taxes on equipment sales and water activities,
vi) Hotel room surcharges, and
vii) National and international conservation organizations.

Transparency

The administration of funding should be clear and transparent to all partners providing the financing. Transparency must be paramount in the use of financial resources to prevent corruption and misunderstandings when handling large sums of money. The creation of steering committees, project oversight committees, or similar groups with stakeholder participation (e.g., environmental managers, researchers, providers of tourist services) is useful and recommended.

Appendix 1. Communication and Outreach Strategies

The following are lionfish-specific communication and outreach strategies that have been successfully implemented within the region. The list is divided into Private, Public and Government sectors.

Private Sector

i) Conduct workshops and trainings (dive industry, fishermen, tourism associations).
ii) Design and disseminate posters, stickers and printed materials (all sectors),
iii) Target and publish in select industry publications,
iv) Request funding from private companies and educate in the process,
v) Give presentations at trade association gatherings,
vi) Maintain reporting 'hotlines' and email (all sectors),
vii) Stress financial benefits of control,
viii) Target private sector at a regional scale,
ix) Target airports and seaports,
x) Promote recognition of partners (all sectors),
xi) Target the food and hospitality sector, and
xii) Establish a regional 'Day of the Lionfish'

Government

i) Disseminate results of research,
ii) Use research data and analysis results to motivate government action,
iii) Prepare response plan prior to invasion,
iv) Use response plan to motivate government action,
v) Train and mobilize first responders,
vi) Create network of regional bodies and governments,
vii) Be persistent in communication to key political leaders,
viii) Target key government meetings to promote inter-governmental coordination,
ix) Conduct face to face meetings with government officials,

x) Utilize other stakeholder groups to engage government officials (e.g., conservation NGOs),

xi) Create policy briefs,

xii) Invite officials to events and to see lionfish in the water,

xiii) Communicate an ecological and economic message to influence government officials, and

xiv) Aim E&O at significant government leaders and be ambitious.

Public Sector (including social/civic organizations and NGOs)

i) Engage the media with press releases and public service announcements,

ii) Host workshops and lectures for the public,

iii) Develop school-based outreach materials,

iv) Recruit a high level or famous individual as a spokesperson,

v) Build broad stakeholder involvement,

vi) Create and host tournaments and fishing derbies,

vii) Engage schools and universities,

viii) Standardize informational materials for all sectors,

ix) Use the internet and other available technology (e.g., social networking sites),

x) Host 'Lionfish Tasting' events,

xi) Host information fairs,

xii) Create displays for aquariums and museums, and

xiii) Target public health issues and hospitals.

Appendix 2. NOAA/CCFHR Venomous Fish Handling Protocol

1. PURPOSE: Establish specific procedures and training requirements for NOAA divers assigned to the *Center for Coastal Fisheries and Habitat Research (CCFHR)* when diving where encounters with venomous fish are likely to occur

2. BACKGROUND: Venomous fish are commonly found in waters of the Southeastern United States, the Gulf of Mexico, and US territories in the Caribbean Sea. NOAA Scientific Divers are increasingly being asked to capture venomous fish, specifically the invasive Indo-Pacific lionfish, for research aimed at understanding and potentially controlling the rapid spread of these species.

3. SCOPE: This document provides operational guidance to CCFHR divers when capturing or handling venomous fish in the water when using SCUBA or while skin diving.

4. POLICY: All divers diving on a CCFHR project shall use protective gear and equipment when working with and around venomous fish.

   Responsibility

   4.1 *Diver* — Each diver shall evaluate his or her risk and complete training prior to conducting dives to collect venomous fish.

   4.2 *Divemaster (DM)* — The DM shall brief the boat crew, divers, and topside support personnel on the risks of handling venomous fish and basic first aid for venomous fish incidents. Further, the DM shall verify that each diver has the appropriate personal protective equipment (PPE) available and agrees to use their PPE prior to allowing dives to commence.

   4.3 *Unit Diving Supervisor (UDS)* — The UDS shall ensure DMs and Divers have the appropriate training, PPE, and appropriate first aid supplies at the dive site prior to authorizing dives.

5. GUIDANCE:

   5.1 *Personal Protective Equipment (PPE)* — Whether capturing live or spearing venomous fish, divers should wear at least one **puncture proof glove** and **full wet suits.** Some divers wear a puncture proof glove on one hand and a regular wet suit glove on the hand they use to control their spear. The buddy diver carrying the **catch bag** shall wear 2 puncture proof gloves. A wetsuit reduces risk but fish spines can penetrate a wetsuit. Thick, clear plastic bags made for use as dry bags are preferred to opaque catch bags, because solid materials resist punctures and because they are clear the diver can observe the fish (Figure 1), and because they might contain some of the fish scent. Canvas material is sometimes used but does not offer the advantages of the heavy clear plastic bags. There is a list of places to acquire protective gear in the Resources Section following the Figures below.

5.2 *Capturing Methods* — Most stings occur when a diver is distracted and not watching the fish when it is captured or bagged, when allowing the bag to hang near the body or leg, or when not wearing proper gloves, or trying to handle a fish by grasping the body. Divers should constantly monitor the location of the bag relative to themselves and other nearby divers, the condition of the bag, the fish in the bag, and the fish being hunted (maintain situational awareness). The dive boat should provide a weighted hang line on which to clip the bag near the safety stop so the divers can finish the ascent unencumbered. When "live boat" diving, divers should hang the catch bag from a line attached to a surface marker buoy or lift bag. A small weight can be added to the catch bag to ensure the bag remains vertical and doesn't drift up during ascent when the fish bladders swell and the fish lose buoyancy control. Good buoyancy control reduces the likelihood of a leg or torso puncture (Figure 2).

5.3 *Live captures* — Nets can be used effectively to capture venomous fish. Most venomous species do not seem to fear divers and do not hurry away from one or two careful attempts to capture them. Closed, clear plastic nets have been effectively used for smaller fish and mesh landing nets for larger specimens. The following 2 different methods are being used:

5.3.1 — For smaller fish, one commonly used strategy is to equip one diver with two nets (one for herding and one for collecting (Figure 3) and the other diver with a collecting bag. Once the first diver captures a fish between the nets, both divers should move to a suitable area to transfer the fish. The preferred area will be protected from strong currents and allow space for equipment and bags without disturbing the nearby benthic habitat. The (bagger) diver with the collection bag places the bag on the bottom, opens the bag and vents any remaining air from the bag. The diver with the netted fish should place both nets on the bottom, collapse the nets to prevent movement of the fish, firmly gasp the fish by the head (and away from the spines- Figures 4 & 5) from the outside of the net, then turn the net inside out and place the fish into the open bag. The diver with the bag should then close the entrance of the bag around the diver's arm who will then release the fish into the bag. The bag can then be closed and secured. To introduce additional fish into the bag, herd and secure previously captured fish towards the bottom of the bag prior to adding the new fish.

5.3.2 — Another commonly used method is as follows: Net the fish, trap it by overturning the net onto the bottom, grasp the cod end of the net (on the thick mil plastic right above the mesh) and with the other hand grasp the net at the opening end trapping the fish between both hands. The bagger then opens up the collection bag, which is held vertically, and the collector the puts the opening of the bag over the top of the collection bag and releases the hand closest to the collection bag. Use the hand closest to the cod end to push forward and move the fish into the bag. The fish want to swim down escape by swimming up is unlikely.

5.4 *Spear captures* — Divers should use spears with paralyzer tips to immobilize the fish when speared (Figure 4). Once speared, the fish should be removed by grasping the fish by the head, away from the venomous spines (Figure 5). Place the fish into collection bags in the same procedure as with nets. Collection bags may also be equipped with a trap door so the fish can be pulled off the spear when placed in the bag. Bags equipped this way reduce handling time and risk. Some divers pith the fish after removing if from the spear tip to immobilize it. To pith a fish, press the tip of a spear point (or a tool like a marlinspike or ice pick) into the head behind the eyes through the brain.

5.5 *Other Safeguards* — Each diver and the DM should assess the potential for and presence of predators, availability of proper gear, the planned methods, each diver's ability and experience. For any method, divers should agree on the method before diving, practice the method in "dry runs" on the surface, and consult with more experienced divers for additional recommendations. No diver should agree to dive unless he or she is comfortable performing the required tasks. The DM should not allow dives to commence unless he or she is absolutely comfortable the dives can proceed without unnecessary risk.

6. FIRST AID:

6.1 *Signs and Symptoms* — Signs and symptoms vary from mild swelling and pain to tachycardia, hypertension, hypotension, seizures, chest pain, abdominal pain, sub dermal necrosis at the sting site, and temporary paralysis to the extremities. Symptoms will vary depending on the severity of the sting. For an average sting, the pain, which may be excruciating, will usually lessen after a few hours. Generalized symptoms can be severe. The distress caused by the pain may advance to a state of delirium in severe cases. The generalized weakness that may develop can involve the cranial nerves, facial muscles, vision, speech, and cause respiratory distress. If stung, the diver should ascend immediately following normal procedures.

6.2 *Treatment*: An envenomated diver should ascend without delay — continuing the dive places the diver at greater risk that a severe reaction would occur while still in the water.

As soon as practical after the diver is out of the water, apply heat to the affected area, either by soaking the affected part in non-scalding hot water up to 45° C (113 °F) or by using a heat compress for 30-90 minutes or until the pain no longer recurs when removed from the water. Avoid burning the victim, who may not be able to detect burn causing temperatures above 120 °F. The wound site should be cleaned with soap and water or rinsed with sterile saline. If a spine or spines have broken off the fish and are still lodged in the skin, it should be gently extracted. Small amounts of bleeding may help to flush some of the venom, but profuse bleeding (rare in these cases) should immediately be controlled. Once heat therapy is ended, the extremity should be immobilized and elevated.

If the victim shows signs of an allergic reaction, simple medications containing anti-histamines may be administered if the victim has no allergies to the medications and they are not nauseated or showing signs of shock. If an epinephrine injector (epi-pen) is onboard, and a person qualified in the use of an epi-pen recommends it be used, an injection might benefit the injured diver with a severe allergic reaction.

Monitor for signs of shock. If the victim's respirations become weak, rapid, or labored or their pulse becomes weak or rapid, immediately initiate the Dive Emergency Assistance Plan (DEAP), provide 100% Oxygen, and be prepared to transport the victim to advanced medical care.

7. OTHER INFORMATION

7.1 Antivenins may be available for some venomous species in areas where these species and humans interact with some frequency. Many proteinaceous toxins, including those of the Indo-Pacific lionfish species, become less potent after ice or heat are applied. However, when cleaning fish precautions should always be taken to prevent punctures since other toxins, allergens, or pathogens might also be present.

7.2 In all cases, envenomations require cessation of future dive activities until the consulting medical practitioner indicates the medical emergency is over and the diver is fit to resume diving.

RESOURCES:

http://www.REEF.org/catalog/83/field_supplies
http://www.jblspearguns.com/polespears.php?id=Polespears
http://www.hexarmor.com/technology/
http://traveloasis.com/glaccleardry.html
http://www.westmarine.com/1/3/west-marine-dry-bags
(Mention of trade names or commercial products does not constitute endorsement by the United States Government.)

Figure 1: Collecting Bag.

Figure 2. Good Buoyancy Control.

Figure 3. Use of 2 Capture Nets.

Figure 4. Pole Spear or Hawaiian Sling with Paralyzer Tip.

Figure 5. Removing an Indo-Pacific Lionfish from a Pole Spear.

Appendix 3. Example of a liability release used in lionfish derbies or other collecting events from the Reef Environmental Education Foundation

Lionfish Biology and Venomology Understanding of Risk

Overview

Lionfish (*Pterois volitans/miles*) are members of the scorpionfish family native to the Indo-Pacific. They are now also found in the waters throughout the Western Atlantic from Rhode Island through the Bahamas Caribbean and Gulf of Mexico. Lionfish are predators reaching a size of more than 470 mm and reproducing year-round in the southern portion of their Atlantic range. They also possess venomous, spines and potentially represent both ecological and human health risks. Significant study, public education/outreach and control efforts are now being planned and implemented throughout the region.

Need for specimens

In order to predict the spread and impact of this invasive species, data are needed to determine populations and distribution, age and growth, reproduction, predation, larval disbursement, recruitment, mortality, genetics and changes to native fish communities. Obtaining this information requires the collection of many samples and appropriate preservation of these samples. There is little funding and few field operations available to collect the number of samples required for statistically robust analysis. REEF is working with U.S. state, federal, and university researchers in addition to Caribbean nations to obtain the necessary samples as well as educate the public and key government officials about the lionfish issue. As part of this effort, we are also collaborating with the dive community to gather samples and data from their regularly dived sites.

Venemology

Lionfish spines, like those of many other scorpionfishes, contain venom. Spines located within the dorsal, ventral, and anal fins contain venomous tissue along the length of the spine. A sheath or skin covers the spine and glandular tissue which, when exposed in the act of puncture, releases venom into the wound. Severities and reactions to stings can vary however lionfish stings are rarely lethal. Reactions to venom typically include localized, moderate to severe, pain and swelling, which normally subsides within hours and with first-aid treatment. Complications or increased severity of reactions could result from allergic reactions or severe punctures. Heat and time reduce the effect of this venom. As with any wound, cleaning and sterilization of the affected area is important to prevent infection. If stung, seek medical attention immediately.

Participation

While the likelihood of coming in contact with a lionfish in the water or on the vessel is extremely small for those wanting to avoid contact, there never-the-less exists the possibility of being stung. No participants wishing to avoid collection, dissection or any other physical contact with lionfish will be asked to handle or in any other way physically engage themselves with any lionfish. All contact with lionfish, direct or otherwise, is voluntary in nature as is participation in this project, knowing the venomous nature of the subject.

I understand these risks and am willing to take part in this project anyway.

Printed Name of Participant: ______________________________

Signed: ______________________________

Date: ______________________________

Signature and date of legal guardian required if participant is under 18.

Legal Guardian Printed Name: ______________________________

Signed: ______________________________

Date: ______________________________

Appendix 4. Sample Datasheet

This sample can be amended for various types of collectors or events

Researcher information*

*Personal information is your choice to provide, it is used for our records only. It will not be distributed. This also helps us distribute collector awards and incentive prizes.

Name: ____________

Address: ____________

____

Phone/email ____________

Are you a (**pick 1**):

Recreational diver/snorkeler   Rec. Fisher
Commercial diver/snorkeler   Comm. Fisher
Other ________(describe)   Researcher

Site Information **complete to the best of your ability.

Date: ___/___/___ mon. / day / year

Site Name: ____________

Latitude (N) ____________

Longitude(W) ____________

Water Temp(F): ______ Depth (ft): ______

Nearest Island/State/Country: ____________

Gear used & effort (**pick 1**)

SCUBA w/net   Snorkel w/net
SCUBA w/spear   Snorkel w/spear
______/______ (# divers / time (hrs) per diver)
Hook & Line____/____(#anglers/time (hrs))
Seine________/_____ (size/# hauls)
Trap _____/___/____ (type/total #/soak t (hrs))
Trawl________/_____ (size/time (min))

Surrounding Habitat (**pick 1**):

Artificial wreck/reef   Mud
Beach   Mangrove
Coral reef   Patch Reef
Cave   Rocks
Hardbottom   Sand
Live rock – reef   Seagrass
Manmade/bridge/piling   Unknown

Please provide as much detail for each fish as possible. Use addition room on back or other sheets as needed for fish from this collection.

FishID Your choice here is an ex.: (stateMonth-year-collector#)	Fish Measurements: (weight in **grams**, all other measurements in **millimeters**)			Tissues Collected (check all that apply)					Gender (Male, Fem., Unk.)	Fat vol. (mL)	Notes
	Weight (grams)	Total L. (mm)	Std. L. (mm)	Otolith	Fin Clip	Gill Clip	Muscle	stomach			
FL3-20-bjon1 (ex: Florida, March- 2020- bJones)	52g	171	130	✓	✓	✓		✓	M		Stomach empty

For electronic data files send to James.Morris@noaa.gov, please make note of the corresponding electronic file name on this page.

Appendix 5. Lionfish Medical Questionnaire

Bahamas National Lionfish Response Project

TREATMENT OF LIONFISH ENVENOMATION QUESTIONNAIRE

HEALTH CARE PROFESSIONAL INFORMATION

1. Specific occupation of healthcare professional:
   a) Physician
   b) Nurse
   c) Other (please specify)
2. Type of healthcare facility:
   a) Public hospital
   b) Private hospital
   c) Government community clinic
   d) Private doctor's office

3. Name of healthcare facility:______________________________

4. Location of healthcare facility: ______________________________

PATIENT INFORMATION

5. Date of healthcare facility visit (dd/mm/yy): ____________________
6. Gender: Male Female
7. Age: <5yrs 5-11yrs 12-25yrs 26-35yrs 36-45yrs 46-55yrs >56yrs
8. Significant medical history:
   - Diabetes Yes No
   - High Blood Pressure Yes No
   - High Cholesterol Yes No
   - Anemia Yes No
   - Low Blood Pressure Yes No
   - Other (please specify)
9. Has the patient ever been envenomated by a lionfish prior to this incident? Yes No
10. When was the last time the patient received a vaccination against tetanus?
   a) Never
   b) Within the last 5 years
   c) Within the last 10 years
   d) More than 10 years
   e) Don't know

INCIDENT REPORT

11. Date of lionfish envenomation (dd/mm/yy): ______________________________
12. Activity at time of envenomation:
   a) Swimming
   b) Fishing
   c) SCUBA Diving
   d) Other (please explain) ______________________________

Bahamas National Lionfish Response Project

13. How much time transpired between envenomation and patient's decision to seek professional help? (Circle one):

a) < 1 hour b.) hours
b) Days d) weeks

14. How much time transpired between envenomation and the patient actually obtaining help from a healthcare professional?

a) Hours
b) Days
c) Weeks

15. Did the patient know the first aid response to lionfish envenomation at the time of the incident? Yes No

16. Did the patient receive first aid prior to seeking professional help? Yes No

17. If "yes" to question 16, what was done? (Circle those that apply):

a) Immersed the wound in hot water

Appendix 6. The Bahamas' Response

THE BAHAMAS' RESPONSE TO THE LIONFISH INVASION

The Department of Marine Resources
Nassau, Bahamas
31 May 2011

Introduction

The Bahamas and the rest of the Wider-Caribbean have been faced with the challenge of combating the Indo-Pacific lionfish (*Pterois volitans*). The lionfish has become a major cause for concern for a number of reasons:

i) Threat to human health;
ii) Ability to out-compete native species due to their ambush style of predation;
iii) Preys on a wide variety of species including several that are commercially important; and
iv) Few known predators in the region.

In 2009, the Department of Marine Resources in collaboration with The College of The Bahamas Marine and Environmental Studies Institute (COB-MESI) initiated a long-term National Lionfish Response Plan (NLRP). The NLRP was formalized in response to the growing threat posed by the invasion of lionfish in the waters of The Bahamas.

The existence of the NLRP is due in part to the mandates set-forth in the National Invasive Species Strategy for The Bahamas (NISS). The National Invasive Species Strategy is essential to this discussion because it laid the framework for proposed amendments to fisheries regulations. Since the drafting of the NISS and subsequently the NLRP, recommendations to amend several regulations in Chapter 244 of the Fisheries Resources (Conservation and Jurisdiction) Regulations have been proposed by The Department of Marine Resources.

National Invasive Species Strategy

The National Invasive Species Strategy (NISS) was published in 2003 with the intent of assessing *current mechanisms existing in the Bahamas to address the invasive species issue*. The NISS outlines an overview of invasive species; identifies mechanisms for the management and control of invasives; and provides recommendations for action. Of particular interest to this discussion are **Recommendations 5, 7, & 8.**T hese recommendations laid the groundwork for the NLRP that would come six (6) years later.

Recommendation 5

Existing Legislation related to management and control of alien species should be enforced and where deficient, be amended. There will also be a need to ***draft and amend new legislation specific to management and control of invasive alien species.***

Recommendation 7
Priority species should be listed for eradication and control. These lists therefore would not include all known invasives for The Bahamas . . . ***species on the eradication and control lists would be reviewed on a regular basis with the result that. . . new species added.***

Recommendation 8
The Government accept and implement an Invasive Species Policy...

The NISS also includes a *Draft National Invasive Species Policy* to be adopted by the Government of the Commonwealth of The Bahamas. The Draft Policy set-forth specific action items for the Government of The Bahamas to carry out. Several of which echo the sentiments of Recommendations 5, 7 & 8:

i) Enact legislation to prevent the introduction of, to control, and to eradicate those alien invasive species which threaten the ecosystems, habitats, endemic species and the human health and welfare of The Bahamas, in support of the Convention on Biological Diversity;
ii) Prepare a National Invasive Species Strategy for The Bahamas, which lists and prioritizes in order of significant impact those invasive species present in the Bahamas;
iii) *Prepare Strategic Management Plans for individual species of high priority as identified under the NISS*;
iv) Conduct and facilitate research into the best management and control practices for individual species, including plants, animals and micro-organisms, using chemical, physical and biological methods that are environmentally sound;
v) Mandate cooperation between Government Ministries, Departments and other Agencies including Non-Governmental Organizations and the Private Sector, as necessary, to implement this policy and carry out the Strategy.

In addition, the NISS outlines codes of conduct for The Government to follow. In particular, The Government is asked to *enforce existing invasive species legislation at all levels, and* ***enact new legislation where deficiencies occur in existing legislation***. This mandate paved the way for the development of the National Lionfish Response Plan and amendments to the Fisheries Resources (Jurisdiction and Conservation) Regulations Chapter 244.

However, when the NISS was first developed, lionfish were not identified as a species recommended for eradication or control, and for good reason. It was not until 2004 that the first sighting of lionfish in The Bahamas was recorded. Currently, the NISS is being updated to incorporate the management of lionfish under the ***Mitigating the Threats of Invasive Alien Species in the Insular Caribbean (MTIASIC) Project***.

The National Lionfish Response Plan
The National Lionfish Response Plan (NLRP) was designed to be a *Strategic Management Plan.* It was determined that the NLRP would function to maintain the distinctiveness and diversity of Bahamian marine communities, protect commercially important fisheries, and safeguard public health by controlling and

The main objectives of the NLRP are to build research; build outreach and educational initiatives; identify sources of financing and other resources; and *develop appropriate invasion policies and regulations via collaborative efforts among local and regional partners in the public and private sector*. These objectives are in accord with recommendations outlined in the NISS. Collectively, they function to ensure the implementation of an efficient lionfish management plan.

Within the response plan, several management goals and strategies have been identified. These goals were selected with the intention of mitigating the threats of the invasive lionfish. They are prioritized in the NLRP by importance: high, medium, and low.

High Priority (immediate action - implementation within 1 - 2 years)

i) Amend existing fisheries regulations:
   A. Allow spear fishing for lionfish within the 1 mile (New Providence) and 200 yard (Family Islands) limit.
   B. Allow an unlimited bag limit per vessel for persons granted a permit to spear lionfish.
   C. Allow the use of SCUBA, spears, and nets to remove lionfish during authorized fishing tournaments.

ii) Encourage the capture and sale of lionfish (for consumption) as a commercial fisheries resource.

iii) Expand and improve the network of National Parks and Marine Reserves.

Medium Priority (implementation within 3 - 5 years)

i) Require the removal of marine debris and artificial structures from near-shore areas.

ii) Improve coastal zone management.

Low Priority (implementation within 6 - 10 years)

i) Ban lionfish as aquarium pets and restrict possession or transport of live lionfish.

The proposed management strategies to amend fisheries regulations are consistent with the code of conduct for the Government of the Bahamas and recommendations in the NISS.

Despite strong support to implement and execute the response plan, the NLRP has not been fully implemented. A cost estimate for the implementation of the plan has not been completed. Aside from the work the Department of Marine Resources is implementing with respect to the NLRP, partner organizations are also contributing the achievement of management goals set out in the plan. The MTIASIC project also seeks to promote the implementation of the NLRP and to essentially formalize the plan after completion of the lionfish control experiments.

Proposed Amendments to Fisheries Resources (Conservation and Jurisdiction) Regulations

In response to recommendations in the NLRP and mandates within the NISS, the Department of Marine Resources proposed amending several regulations in the Fisheries Resources (Conservation and Jurisdiction) Regulations. The main points of the proposed amendments are set out below:

i) Empower the Minister to grant permission for persons to use "prohibited apparatus" for fishing subject to special conditions. This would include SCUBA, hooka and related breathing devices.
ii) Allow fishing for lionfish in areas normally protected from general spearfishing, areas close to coastlines, within harbours
iii) Allow the use of small mesh nets to fish for lionfish,
iv) Allow spearfishing tournaments for lionfish,
v) Allow visitors fishing under the terms of a sportsfishing permit, to collect and take out of the country an unlimited amount of lionfish.

Conclusion

The Bahamas' response to the lionfish invasion has spanned the development of the National Invasive Species Strategy, the National Lionfish Response Plan, and proposed amendments to the Fisheries Resources (Conservation and Jurisdiction) Regulations.

The Bahamas has made some progress in its attempts to address the lionfish invasion and the problem of invasive alien species in general. Considerable progress has been made in respect of building capacity and strengthening relationships between the various partners that have been engaged in this effort. In an effort to continue to move forward, response will be focused in areas such as education, public awareness, and outreach.

Despite these successes, the Bahamas is still faced with various administrative and legal challenges that prohibit it from being able to fully respond to the threat posed by lionfish.

Appendix 7. Special Disposition for Lionfish in Puerto Rico

Fish Regulation 7949
November 2010

ARTICLE 22 - SPECIAL DISPOSITIONS FOR THE LION FISH

22.1 In the case of any species of lionfish (*Pterois volitans, Pterois miles*) allowed:

a. unlimited export agencies. (*further clarified to be export of live lionfish through a permitting system*)
b. the use of juvenile lionfish as bait.
c. fish caught in mangrove areas, lagoons, harbors, estuaries or channels interconnecting water bodies of fresh water with a body of salt water.
d. their catch can be made at any time of day or night.
e. capture without having to comply with the requirements of Control Permit harmful species, as described in Article 19.2 of this regulation (do not need a memorial explanatory resume delivery, limited area, or permits for scientific and educational) .
f. capture in Nature Reserves, Marine Reserves, Special Planning Areas or areas closed to fishing ("no take zones"), only by written authorization of the Secretary for volunteers entitled "Special Authorization Capture Volunteer Lionfish."
g. used as spear fishing gear, "Hawaiian sling" or blunt object or ornamental fish hand nets or other networks approved by the Department to capture lionfish, in combination with "Scuba" or "Hookah" only by written authorization from the Secretary entitled "Special Authorization Capture Volunteer Lionfish."

22.2. In the case of having the "Special Volunteer Lionfish Capture" and use "Scuba" or "Hookah" in combination with Spear, "Hawaiian sling" or blunt object, their capture alone for that day will lionfish.

22.3. In the case of having the "Special Volunteer Lionfish Capture" and use it to fish in Nature Reserves, Marine Reserves, Special Planning Areas or areas closed to fishing ("no take zones"), their capture alone for that day will lionfish.

22.4. Fishing activities or capture lionfish in Nature Reserves, Marine Reserves, Special Planning Areas or areas closed to fishing ("no take zones") empowered by the "Special Volunteer capture lionfish, will have to be reported to the Department by the methods of reporting under the authority of the Secretary. The work will be supervised by the Department, must be understood. The Secretary may regulate the activity of capture by volunteers through Administrative Order.

Appendix 8. Lionfish Response Licenses in the Cayman Islands

CAYMAN ISLANDS DEPARTMENT OF ENVIRONMENT
LIONFISH RESPONSE LICENSE
TERMS AND CONDITIONS

As provided for under PART VI, Section 27.(2) of the Marine Conservation Law (2003 Revision), in issuing a special license to participants of the Lionfish Eradication Program being implemented by the Department of Environment, the Marine Conservation Board has attached the following conditions which shall apply:

1. Only the Red Lionfish (*Pterois volitans*) may be taken under this license.

2. The licensee is exempt from the following prohibitions in the law pertaining to the taking of marine life:
   a) Restriction on the taking of marine life using SCUBA.
   b) Restriction on the use of gloves while on SCUBA.
   c) Restriction on taking marine life within Marine Parks and Replenishment Zones.
   d) Restriction on taking fish less than eight inches in fork length.

3. The following are still prohibited:
   a) Entering the water within the Environmental Zone, Grand Cayman.
   b) Scuba Diving in the No Diving Zones.
   c) The use of any spearing device.
   d) The use of any noxious substances.

4. The licensee must report all catches of lionfish and provide details such as name of culler, date, time, location, number of lionfish and approximate sizes, and dive company (if applicable). For size approximation: S means less than 6 inches, M means 6 to 12 inches, L means over 12 inches. See example below:

John DoE, 1st January 2011, 9:00am, Trinity Caves, 15 lionfish, 2S, 8M, 5L.

5. All sightings of lionfish should be reported.

6. PLEASE HAVE THE LAMINATED LICENSE AVAILABLE WHEN IN THE FIELD COLLECTING LIONFISH. On the boat (or in your car if shore diving) is acceptable.

7. Failure to abide by these conditions will result in immediate revocation of this license and may result in prosecution.

CAYMAN ISLANDS DEPARTMENT OF ENVIRONMENT (DOE) LIONFISH RESPONSE (SPEARGUN) LICENSE

TERMS AND CONDITIONS FOR DIVE COMPANIES

As provided for under PART VI, Section 27.(2) of the Marine Conservation Law (2003 Revision), in issuing a special license to participants of the Lionfish Eradication Program being implemented by the Department of Environment, the Marine Conservation Board has attached the following conditions which shall apply:

1. Only the Red Lionfish (*Pterois volitans*) may be taken under this license.

2. The licensee is exempt from the following provisions in the law pertaining to the taking of marine life :
    - A. Restriction on the taking of marine life using SCUBA.
    - B. Restriction on the use of gloves while on SCUBA.
    - C. Restriction on taking marine life within Marine Parks and Replenishment Zones.
    - D. Restriction on taking fish less than eight inches in fork length.
    - E. The use of a spearing device provided that the following conditions are adhered to:
        - a. Only DoE issued spearguns may be used under this license
        - b. Only licensed Lionfish Cullers currently employed by licensee may use the licensed spearguns
        - c. Each speargun will be tagged and licensed to one company but will remain property of DoE
        - d. The licensee shall not lend, give, or sell their spear gun to any other person.
        - e. License must accompany speargun at all times
        - f. Each company will provide DoE with a complete list of all employees for approval to use the spearguns
        - g. Spearguns must be kept in a secure location at place of employment
        - h. Companies must notify DoE if approved employees leave any lost spearguns or broken tags must be reported to DoE IMMEDIATELY
        - i. DoE will not be held responsible for any injuries resulting from the use of these spearguns

3. The following are still prohibited:
    - A. Entering the water within the Environmental Zone, Grand Cayman.
    - B. Scuba Diving in the No Diving Zones.
    - C. The use of any noxious substances.

4. The licensee must report all catches of lionfish and provide details such as name of culler, date, time, location, number of lionfish and approximate sizes and dive company (if applicable).

For size approximation:

S means less than 6 inches, M means 6 to 12 inches, L means over 12 inches. See example below:

John DoE, 1st January 2011, 9:00am, Trinity Caves, 15 lionfish, 2S, 8M, 5L.

5. PLEASE HAVE THE LAMINATED LICENSE AVAILABLE WHEN IN THE FIELD COLLECTING LIONFISH. On the boat (or in your car if shore diving) is acceptable.

6. Failure to abide by these conditions will result in immediate revocation of this license and may result in prosecution.

7. The Marine Conservation Board retains the right in its discretion to revoke any license.

CAYMAN ISLANDS DEPARTMENT OF ENVIRONMENT (DOE)
LIONFISH RESPONSE (SPEAR GUN) LICENSE

TERMS AND CONDITIONS FOR RESIDENTS

As provided for under PART VI, Section 27.(2) of the Marine Conservation Law (2003 Revision), in issuing a special license to participants of the Lionfish Eradication Program being implemented by the Department of Environment, the Marine Conservation Board has attached the following conditions which shall apply:

1. Only the Red Lionfish (*Pterois volitans*) may be taken under this license.

2. The licensee is exempt from the following provisions in the law pertaining to the taking of marine life :
   A. Restriction on the taking of marine life using SCUBA.
   B. Restriction on the use of gloves while on SCUBA.
   C. Restriction on taking marine life within Marine Parks and Replenishment Zones.
   D. Restriction on taking fish less than eight inches in fork length.
   E. The use of a spearing device provided that the following conditions are adhered to:
      a. Only DoE issued spear guns may be used under this license
      b. Only licensed Lionfish Cullers may use the licensed spear guns
      c. Each spear gun will be tagged and licensed to one individual and may only be used by that person but will remain property of DoE.
      d. The licensee shall not lend, give, or sell their spear gun to any other person.
      e. License must accompany spear gun at all times
      f. Spear guns must be kept in a secure location.
      g. Any lost spear guns or broken tags must be reported to DoE IMMEDIATELY
      h. DoE will not be held responsible for any injuries resulting from the use of these spear guns

3. The following are still prohibited:
   A. Entering the water within the Environmental Zone, Grand Cayman.
   B. Scuba Diving in the No Diving Zones.
   C. The use of any noxious substances.

4. The licensee must report all catches of lionfish and provide details such as name of culler, date, time, location, number of lionfish and approximate sizes and dive company (if applicable). For size approximation:

S means less than 6 inches, M means 6 to 12 inches, L means over 12 inches. See example below:

John DoE, 1st January 2011, 9:00am, Trinity Caves, 15 lionfish, 2S, 8M, 5L.

5. PLEASE HAVE THE LAMINATED LICENSE AVAILABLE WHEN IN THE FIELD COLLECTING LIONFISH. On the boat (or in your car if shore diving) is acceptable.

6. Failure to abide by these conditions will result in immediate revocation of this license and may result in prosecution.

7. The Marine Conservation Board retains the right in its discretion to revoke any license.

Appendix 9. Disclaimer for Scuba Divers Capturing Lionfish in Cozumel, Mexico

Diving with SCUBA
Disclaimer of Responsibilities and Risks

Please read carefully and fill the blanks before signing.

I_______________________________, hereby say that I am a certified diver and I understand the dangers inherent in the diving with SCUBA (Self Contained Underwater Breathing Apparatus). I also understand that these risks include, but are not limited to damage lung expansion, drowning, decompression sickness and other damage that may occur in the sea. To sign this waiver of responsibility, I certify that I was warned and fully informed and therefore expressly assume these and other risks during my tenure as a “volunteer” in the “lionfish capture and control program” promoted by the “Parque Nacional Arrecifes de Cozumel” (CONANP).

I understand and agree that any member of staff of the “Parque Nacional Arrecifes de Cozumel”, none of its employees, officers, agents, or respective assignees hereinafter calls ("exonerated parties") be considered liable or responsible in any way for any injury, death or other damage suffered I or my family, heirs or assignees that may occur as a result of my participation in the dives with scuba.

In addition, I affirm that I have legal age and I am trained by law to sign this waiver of responsibilities or that I have obtained consent in writing of my parent or guardian.

I _______________________________ USING THIS INSTRUMENT RELEASED AND WAIVE TO ALL THE ABOVE MENTIONED, INSTITUTIONS AND/OR INDIVIDUALS FROM ALL LIABILITY FOR PERSONAL INJURY, DAMAGE TO PROPERTY OR DEATH BY NEGLIGENCE IN ANY WAY EXCLUDING BUT NOT LIMITED TO NEGLIGENCE OF EXEMPTED PARTIES, EITHER PASSIVE OR ACTIVE.

I_______________________________ AM IN KNOWLEDGE AND I HAVE READ THE PREVIOUS PARAGRAPH, I COMPLETELY UNDERSTAND THE POTENTIAL DANGERS THAT EXIST IN THE DIVES WITH SCUBA, I AM QUITE AWARE LEGAL CONSEQUENCES AND I AM IN AGREEMENT IN WHICH THIS DOCUMENT IS LEGALLY CONFORMED AND ME DISABLES ENGAGE ECONOMIC DEMAND AGAINST PREVIOUSLY LISTED, ENTITIES OR INDIVIDUALS, ARE SPECIFICALLY NAMED OR NOT, FOR PERSONAL INJURY, DAMAGE TO PROPERTY OR DEATH BY NEGLIGENCE OR PRODUCT RESPONSIBILITIES.

NAME _______________________________

SIGNATURE _______________________________ DATE _______________

Appendix 10. Brief Summary of Existing Regulations Regarding IAS in the French West Indies (and French Guiana)

Given the status of departments of Guadeloupe and Martinique within the Republic of France, the general framework for IAS is the French legislation and corresponding regulations, with local specifications or adaptations. In France the legislation on marine and terrestrial wild species is two-fold :

i) *Protected species.* Some species are given a particular status of protected species, meaning that they cannot be killed, caught, harassed, etc.

   Several regimes co-exist: species protected on the entire national territory (e.g whales, strictly protected in all French waters all around the world) ; species protected only in some locations by local decree, whether this particular status is justified by their restricted occurrence -it would make little sense to protect in European France some species that are endemic to Guadeloupe - or by a specially fragile local status whereas the national status of the species is favourable.

ii) *« Harvested » species* — These are the species that can be hunted, fished, or more generally taken or caught. Regarding fisheries, quotas (total catches authorized) are determined for all fished species at the national level in accordance with European decisions. These national quotas can be complemented locally by local specific decrees that regulate fishing periods (open/close seasons), minimal/ maximal sizes of fishes caught, or that can edict particular prohibitions of fishing for some species in particular when there is a pollution or a risk of contamination with ciguatera.

All this implies that, in Martinique and in Guadeloupe, the quotas determined nationally apply, completed by specific decisions taken at the local level that add particular regulations or framework

Noticeably, all fishing devices are strictly forbidden when scuba diving ; fishing is only permitted for snorkellers.

Invasive alien species in general

Introduction pathways: France has ratified in 2008 the IMO Convention on Ballast Waters.

A recent report prepared by the French IUCN committee summarizes the existing regulations and their gaps regarding the control and destruction of IAS in France (including Guadeloupe and Martinique). The legal possibility to take control measures depends on the legal status of the target species :

i) If it is protected, of course it cannot be destroyed until the protected status has been removed;
ii) If it the species is domestic or a crop (not a wild species), then the situation is clear and well framed by regulatory texts;
iii) If the species is wild and not protected, then it can be difficult to know exactly what measures can be taken against a particular species. The complexity arises noticeably from the scattering of legal articles between several laws and codes, which hampers having a clear, harmonious understanding at once. Furthermore several state departments and agencies often have shared competences on the issue, collaboration is thus required before measures can be taken; however the situation is more simple within protected areas.

The main gaps, conflicts or confusions occurs with wild IAS animal species that have not been given the status of « pest » or nuisance, the definition of this status being most often made in relation with agriculture or farming. Indeed if an animal IAS species has no negative consequence on crops or domestic species, then it is less easy to confer it the status of nuisance. This is a problem because once a species is declared a nuisance, the situation is much more clear and control measures (and corresponding funding) are far easier to design, agree upon and implement.

For wild IAS species that have no consequences on agriculture or farming, the Environment Code is the reference in terms of control. This code allows the administration to capture, take, keep or destroy the specimens of the introduced species, as soon as its presence in the wild is documented. All IAS species are covered by this provision, but its enforcement is restricted to nominative lists of species decided by the Minister of Environment. It means that, in Guadeloupe or in Martinique, the services in charge of environment have not the mandate to decide on their own that the lionfish is to be put on the above-mentioned list: they can propose it, but the decision has to be made by the ministry in Paris.

For the lionfish

As explained before, a species has to be put by the ministry on a national list before special decision for its destruction can be made. This is the case for the lionfish : it has not yet be listed in the national IAS list in France, henceforth no particular regulation exist on the lionfish at the national level. The consequences are:

i) That the extent to which local decisions to control it can be made is limited (special use of fishing gears, prohibition of sale of living specimens, etc.). **A local decree has been passed some months ago in Guadeloupe authorizing the capture of lionfish for scuba divers with gun or special kit. The authorization is granted only for a list of people whose names are listed in the decree. A similar decree has been passed in Martinique**, but as long as the lionfish is not on the national IAS list, both decrees have to rely on a nominative and limited list of individuals, meaning that no global control campaign can be launched easily, except with authorized fishing gears (nets etc) but not with scuba divers.

ii) Incidentally, nothing prohibits today the fishing and trade of the lionfish in France, except the restriction for scuba diving and its exemption mentioned above. This could be questioned as the risk of contamination with ciguatera is not known yet

An important step now is to try and have the lionfish listed by the ministry as an IAS in the FWI so that a ministerial decree authorizing its destruction can be signed, with corresponding funding granted. The local state delegations for environment in Guadeloupe and Martinique are working on that.

Appendix 11. Excerpt from Bonaire Marine Park Island Resolution (AB2010 #14)

Section IV

ARTICLES CONCERNING FISHING
Article 9

i) It is forbidden to use mechanical gear, explosives, hand spears or poles with hooks, to hunt or catch marine life.
ii) It is forbidden to transport, to offer for sale, to transfer, or to deliver marine life that has been caught by mechanical marine hunting gear, explosives, hand spears or hand hooks.
iii) It is forbidden to transport mechanical marine hunting gear over public ground or public water.
iv) It is forbidden to offer for sale, to sell, or to have in stock for sale in a shop or accompanying space, mechanical marine hunting gear complete or in parts.
v) Mechanical marine hunting gear means any guns and pistols that are regulated by the Firearms Ordinance 1931 (P.B. 1931 nr. 2) as amended, as well as any guns or pistols that either by air or by gas under pressure, or by other means, can fire missiles under water.

With the exception of explosives, the prohibitions set forth in paragraph 1, 2 and 3 do not apply to the Manager or to the persons assigned by the Manager, upon catching, gathering and killing of harmful species as referenced in paragraph 1, Article 19 of the Island Resolution Nature Management Bonaire

LITERATURE CITED

Ahrenholz, D.W. and J.A. Morris, Jr. 2010. Larval duration of the lionfish, *Pterois volitans,* collected from the Bahamian Archipelago. *Environmental Biology of Fishes* **88**(4): 305-309.

Albins, M.A. 2012. Effects of the Invasive Pacific Red Lionfish *Pterois volitans on Native Atlantic Coral-reef Fish Communities.* Ph.D. Dissertation. Department of Zoology, Oregon State University, Corvalis, Oregon. Available: http://ir.library.oregonstate.edu/xmlui/handle/1957/24859. Last access date: July 31, 2012.

Albins, M.A. 2012. Effects of invasive Pacific red lionfish (*Pterois volitans)* vs. a native predator on Bahamian coral-reef fish communities. *Biological Invasions* DOI 10.1007/s10530-012-0266-1.

Albins, M.A. and M.A. Hixon. 2008. Invasive Indo-Pacific lionfish (*Pterois volitans)* reduce recruitment of Atlantic coral-reef fishes. *Marine Ecology Progress Series* **367**: 233-238.

Albins, M.A. and M.A. Hixon. 2011. Worst case scenario: potential long-term effects of invasive predatory lionfish (*Pterois volitans*) on Atlantic and Caribbean coral-reef communities. *Environmental Biology of Fishes* DOI: 10.1007/s10641-011-9795-1.

Albins, M.A. and P.J. Lyons. 2012. Invasive red lionfish Pterois volitans blow directed jets of water at prey fish. *Marine Ecology Progress Series* **448**: 1-5.

Badillo, R.B., W. Banner, J.A. Morris, Jr., S.E. Schaeffer. 2012. A Case Study of Lionfish Sting-Induced Paralysis. *AACL Bioflux* **5**:1-3.

Barbour, A., M.L. Montgomery, A.A. Adamson, E. Díaz-Ferguson, and B.R. Silliman. 2010. Mangrove use by the invasive lionfish *Pterois volitans. Marine Ecology Progress Series* **401**: 291-294.

Barbour, A.B., M.S. Allen, T.K. Frazer, and K.D. Sherman. 2011. Evaluating the potential efficacy of invasive lionfish (*Pterois volitans*) removals. *PLoS One* **6**:1-7.

Bath, P. 2011. Fundraising strategies for environmental funds. RedLAC capacity building project for environmental funds. Rio de Janeiro, Brazil. Available: REDLAC http://toolkit.conservationfinance.org/sites/default/files/documents/redlac-capacity-building/redlacfundraising-strategies-environmental-funds08112011.pdf. Last access date: July 31, 2012.

Betancur-R, R., A. Hines, A.P. Acero, G. Orti, A.E. Wilbur, and D.W. Freshwater. 2011. Reconstructing the lionfish invasion: Insights into Greater Caribbean biogeography. *Journal of Biogeography* **38**: 1281-1293.

Bohnsack, J.A. and S.P Bannerot. 1986. A stationary visual census technique for quantitatively assessing community structure of coral reef fishes. NOAA Technical Report NMFS 41. 21 pp.

Bullard, S.A., A.M. Barse, S.S. Curran, and J.A. Morris, Jr. 2011. First record of a digenean from invasive lionfish, *Pterois CF. volitans* (Scorpaeniformes: Scorpaenidae) in the Northwestern Atlantic ocean. *Journal of Parasitology* **97**: 833-8378.

Bunce, L. and B. Pomeroy, (2003) Socioeconomic monitoring guidelines for coastal managers in the Caribbean: Socmon Caribbean. World Commission on Protected Areas and Australian Institute of Marine Science, Australia. Available: http://www.socmon.org/publications.aspx.

Bunce, L., P. Townsley, R. Romeroy, and R. Pollnac (2000) *Socio-economic Manual for Coral Reef Management*. Global Coral Reef Monitoring Network. Available: http://www.socmon.org/pdf/GCRMN_Manual.pdf.

Caldow, C., R. Clark, K. Edwards, S.D. Hile, C. Menza, E. Hickerson, and G. P. Schmahl. (2009). *Biogeographic Characterization of Fish Communities and Associated Benthic Habitats within the Flower Garden Banks National Marine Sanctuary: Sampling Design and Implementation of SCUBA Surveys on the Coral Caps*. NOAA Tech. Memorandum NOS NCCOS 81. Silver Spring, Maryland USA. 134 pp. Available: http://ccma.nos.noaa.gov/products/biogeography/fgb/.

Cesar, H., L. Pet-Soede, M.C. Quibilan, P.M. Alino, H.O. Arceo, I.V. Bacudo, and H. Francisco. 2000. First evaluation of the impacts of the 1998 coral bleaching event to fisheries and tourism in the Philippines. *in* Coral Bleaching: Causes, Consequences and Response. Coastal Management Report #2230. Coastal Resources Centre, University of Rhode Island. ISBN # 1-885454-40-6.

Church, J.E. and W.C. Hodgson. 2002. The pharmacological activity of fish venoms. *Toxicon* **40**: 1083-1093.

Cohen, A.S. and A.J. Olek. 1989. An extract of lionfish (*Pterois volitans)* spine tissue contains acetylcholine and a toxin that affects neuromuscular-transmission. *Toxicon* **27**: 1367-1376.

Colorni, A. and A. Diamant. 2005. Hyperparasitism of trichodinid ciliates on monogenean gill flukes of two marine fish. *Diseases of Aquatic Organisms* **65**: 177-180.

Cowen R.K., C.B. Paris, and A. Srinivasan. 2006. Scaling of connectivity in marine populations. *Science* **311**: 522-526.

Côté, I.M. and S.J. Green. 2012. Potential effects of climate change on a marine invasion: The importance of current context. *Current Zoology* **58**: 1-8.

Côté, I.M. and A. Maljkovic. 2010. Predation rates of Indo-Pacific lionfish on Bahamian coral reefs. *Marine Ecology Progress Series* **404**: 219-225.

Darling, E.S., S.J. Green, J.K. O'Leary, and I.M. Côté. 2011. Indo-Pacific lionfish are larger and more abundant on invaded reefs: A comparison of Kenyan and Bahamian lionfish populations. *Biological Invasions* **13**: 2045-2051.

Diamant, A., C.M. Whipps, and M.L. Kent. 2004. A new species of Sphaeromyxa (Myxosporea: Sphaeromyxina: Sphaeromyxidae) in devil firefish, *Pterois* miles (Scorpaenidae), from the Northern Red Sea: Morphology, ultrastructure, and phylogeny. *Journal of Parasitology* **90**: 1434.

Dojiri, M. and J.S. Ho. 1988. Two Species of Acanthochondria Copepoda Poecilostomatoida parasitic on fishes of Japan. Report of the Sado Marine Biological Station, Niigata University **18**: 47-56.

Freshwater, D.W., A. Hines, S. Parham, A. Wilbur, M. Sabaoun, J. Woodhead, L. Akins, B. Purdy, P.E. Whitfield, and C.B. Paris. 2009. Mitochondrial control region sequence analyses indicate dispersal from the US East Coast as the source of the invasive Indo-Pacific lionfish *Pterois volitans in the Bahamas. Marine Biology* **156**: 1213-1221.

Fishelson, L. 1975. Ethology and reproduction of pteroid fishes found in the Gulf of Agaba (Red Sea), especially *Dendrochirus brachypterus* (Cuvier), (Pteroidae, Teleostei). *Pubblicazioni della Stazione Zoologica di Napoli I* **39**: 635-656.

Fishelson, L. 1997. Experiments and observations on food consumption, growth and starvation in *Dendrochirus brachypterus and Pterois volitans* (Pteroinae, Scorpaenidae). *Environmental Biology of Fishes* **50**: 393-401.

Green, S.J., J.L. Akins, A. Maljkovic and I.M. Côté. 2012a. Invasive lionfish drive Atlantic coral reef fish declines. *PLoS ONE* **7**:DOI:10.1371/journal.pone.0032596.

Green, S.J., Akins, J.L., and J.A. Morris, Jr. 2012b. Lionfish dissection: Techniques and applications. *NOAA Technical Memorandum NOS NCCOS* 139. 24 pp.

Green, S.J., J.L. Akins, and I.M. Côté. 2011. Foraging behaviour and prey consumption in the Indo-Pacific lionfish on Bahamian coral reefs. *Marine Ecology Progress Series* **433**: 159-167.

Green, S.J. and I.M. Côté. 2009. Record densities of Indo-Pacific lionfish on Bahamian coral reefs. *Coral Reefs* **28**: 107.

Grubich, J.R., M.W. Westneat, and C.L. McCord. 2009. Diversity of lionfishes (Pisces: Scorpaenidae) among remote coral reefs of the Palau Archipelago. *Coral Reefs* **28**: 807

Halstead, B., M.J. Chitwood, and F.R. Modglin. 1955. The anatomy of the venom apparatus of the zebrafish, *Pterois volitans* (Linnaeus). *Anatomical Record* **122**: 317-333.

Hamner, R.M., D.W. Freshwater, and P.E. Whitfield. 2007. Mitochondrial cytochrome b analysis reveals two invasive lionfish species with strong founder effects in the western Atlantic. *Journal of Fish Biology* **71**: 214-222.

Hill, J. and C. Wilkinson. 2004. Methods for Ecological Monitoring of Coral Reefs. Australian Institute of Marine Science. Available: http://www.icran.org/pdf/Methods_Ecological_Monitoring.pdf

Imamuara, H. and M. Yabe. 1996. Larval record of a red firefish, *Pterois volitans,* from northwestern Australia (Pisces: Scorpaeniformes). *Bulletin of the Faculty of Fisheries, Hokkaido University* **47**: 41-46.

Jud, Z.R., C.A. Layman, J.A. Lee, and D.A. Arrington. 2011. Recent invasion of a Florida (USA) estuarine system by lionfish *Pterois volitans/P. miles. Aquatic Biology* **13**: 21-26.

Kizer, K.W., H.E. McKinney, and P.S. Auerbach. 1985. Scorpaenidae envenomations: A five-year poison center experience. *Journal of the American Medical Association* **253**: 807-810.

Krebs, C.J. 1999. *Ecological Methodology. 2nd Edition*. Harper & Row, New York, New York USA.

Kulbicki, M., J. Beets, P. Chapanet, K. Cure, E. Darling, S.R. Floeter, R. Galzin, A. Green, M. Harmelin-Vivien, M. Hixon, Y. Letourneur, T. Lison de Loma, T. McClanahan, J. McIlwain, G. MouTham, R. Myers, J.K. O'Leary, S. Planes, L. Vigliola, and L. Wantiez. 2012. Distribution of Indo-Pacific lionfishes *Pterois* spp. in their native ranges: implications for the Atlantic invasion. *Marine Ecology Progress Series* **446**: 189-205.

Labross, P., M. Kulbicki, and J. Ferraris. 2002. Underwater Visual Fish Census Surveys: Proper use and implementation. Secretariat of the Pacific Community, Noumean, New Caledonia. Available: http://www.spc.int/DigitalLibrary/Doc/FAME/Manuals/Labrosse_02_UVC.pdf.

Leis, J.M. and D.S. Rennis. 2000. Scorpaenidae (Scorpionfishes, Stonefishes). Pp. 217-255 in: Leis, J.M and B.M. Carson-Ewart (eds.). *Fauna Malesiana Handbooks. The Larvae of Indo-Pacific Coastal Fishes. An Identification Guide to Marine Fish Larvae.* Brill, Leiden, The Netherlands.

Lesser, M.P. and M. Slattery. 2011. Phase shift to algal dominated communities at mesophotic depths associated with lionfish (*Pterois volitans)* invasion on a Bahamian coral reef. *Biological Invasions* **13**: 1855-1868.

Maljković, A., T.E. Van Leeuwen, and S.N. Cove. 2008. Predation on the invasive red lionfish, *Pterois volitans* (Pisces: Scorpaenidae), by native groupers in the Bahamas. *Coral Reefs* **27**: 501-501.

Mebs, D. and D. Knob. 2004. Rotfeuerfische und ihr Gift. Koralle 25: 40-44.

Menza, C.W., J. Ault, J. Beets, J. Bohnsack, C. Caldow, J. Christensen, A. Friedlander, C. Jeffrey, M. Kendall, J. Luo, M. Monaco, S. Smith and K. Woody. 2006. A Guide to Monitoring Reef Fish In the National Park Service`s South Florida/Caribbean. NOAA Technical Memorandum NOS NCCOS 39. 169 pp. Available: http://www.ccma.nos.noaa.gov/publications/Reefmonitoringguide.pdf.

Mito, S. 1963. Pelagic fish eggs from Japanese waters – III. Percina, VIII. Cottina. IX. Echeneida and Pleuronectida. *Japanese Journal of Ichthyology* **11**: 39-102.

Mito, S. and K. Uchida. 1958. On the egg development and hatched larvae of a scorpaenid fish, *Pterois lunulata* Temminck et Schlegel. *Scientific Bulletin of the Faculty of Agriculture, Kyushu University* **16**: 381-385.

Morris, J.A. Jr. and P.E. Whitfield. 2009. Biology, ecology, control and management of the invasive Indo-Pacific lionfish: An updated integrated assessment. *NOAA Technical Memorandum NOS NCCOS* **99***:* 57 pp.

Morris, J.A., Jr. and J.L. Akins. 2009. Feeding ecology of invasive lionfish (*Pterois volitans)* in the Bahamian archipelago. *Environmental Biology of Fishes* **86**: 389-398.

Morris, J.A., Jr. 2009. *The Biology and Ecology of Invasive Indo-Pacific Lionfish.* Ph.D. Dissertation. North Carolina State University, Raleigh, North Carolina USA.

Morris, J.A. Jr., K.W. Shertzer, and J.A. Rice. 2011. A stage-based matrix population model of invasive lionfish with implications for control. *Biological Invasions* **13**: 7-12.

Moyer, J.T. and M.J. Zaiser. 1981. Social-organization and spawning behavior of the Pteroine fish *Dendrochirus zebra* at Miyake-Jima, Japan. *Japanese Journal of Ichthyology* **28**: 52-69.

Mumby, P.J., A.R. Harborne, and D.R. Brumbaugh. 2011. Grouper as a natural biocontrol of invasive lionfish. *PLoS ONE* **6**: 1-4.

Ogawa, K., M.G. Bondad-Reantaso, and H. Wakabayashi. 1995. Redescription of *Benedenia epinepheli (Yamaguti, 1937) Meserve, 1938* (Monogenea: Capsalidae) from cultured and aquarium marine fishes of Japan. *Canadian Journal of Fisheries and Aquatic Sciences* **52** (Suppl. 1): 62-70.

Osenberg, C. W., R. J. Schmitt, S.J. Holbrook, K.E. Abu-Saba, and R. Flegal. 1994. Detection of environmental impacts: Natural variability, effect size, and power analysis. *Ecological Applications* **4**: 16-30.

Paperna, I. 1972. Monogenea of Red Sea fishes. III. Dactylogyridae from littoral and reef fishes. *Journal of Helminthology* **46**:47.

Paperna, I. 1976. Parasitological survey of fishes of the Red Sea and Indian Ocean. Page 82 in: Reiss, Z. and I. Paperna (eds.). *Fifth Report of the H. Steinitz Marine Biology Laboratory, Elat 1975-1976.*

Rogers, C.S., G. Garrison, R. Grober, Z.M. Hillis, M.A. Franke. 1994. Coral Reef Monitoring Manual for the Caribbean and Western Atlantic. Virgin Islands National Park. US National Park Service. Virgin Islands, USA. 87pp. Available: http://fl.biology.usgs.gov/Monitoring_Manual.pdf

Ruiz-Carus, R., R.E. Matheson, D.E. Roberts and P.E. Whitfield. 2006. The western Pacific red lionfish, *Pterois volitans* (Scorpaenidae), in Florida: Evidence for reproduction and parasitism in the first exotic marine fish established in state waters. *Biological Conservation* **128**:384-390.

Sakai, A.K., F.W. Allendorf, J.S. Holt, D.M. Lodge, J. Molofsky,K. A. With, S. Baughman, R.J. Cabin, J.E. Cohen, N.C. Ellstrand, D.E. McCauley, P. O'Neil, I.M. Parker, J.N. Thompson, and S.G. Weller. 2001. The population biology of invasive species. *Annual Review of Ecology and Systematics* **32**: 305-332.

Saunders, P.R. and P.B. Taylor. 1959. Venom of the lionfish *Pterois volitans. American Journal of Physiology* **197**: 437-440.

Scheiner, S.M. and J. Gurevitch. 2001. *Design and Analysis of Ecological Experiments*. Oxford University Press Inc., New York, New York USA. 432 pp.

Schmitt, E.F. and K.M. Sullivan. 1996. Analysis of a volunteer method for collecting fish presence and abundance data in the Florida Keys. *Bulletin of Marine Science* **59**(2):404-416.

Schofield, P.J. 2009. Geographic extent and chronology of the invasion of non-native lionfish (*Pterois volitans* [Linnaeus 1758] and *P. miles* [Bennett 1828]) in the Western North Atlantic and Caribbean Sea. *Aquatic Invasions* **4**:473-479.

Schofield, P.J. 2010. Update on geographic spread of invasive lionfishes (*Pterois volitans* [Linnaeus 1758] and *P. miles* [Bennett 1828]) in the Western North Atlantic Ocean, Caribbean Sea and Gulf of Mexico. *Aquatic Invasions* **5**: S117-S122.

Schofield, P.J, J.A. Morris, Jr., J.N. Langston, and P.L. Fuller. 2012. *Pterois volitans/ miles* Factsheet. USGS Nonindigenous Aquatic Species Database, Gainesville, Florida USA. Available: http://nas.er.usgs.gov/queries/ FactSheet.aspx?speciesID=963. Last access date: August 1, 2012.

Schultz, E.T. 1986. *Pterois volitans and Pterois miles:* Two valid species. *Copeia* **1986**: 686-690.

Sealey-Baker, M. 2011. Coral Bleaching and Dive Operators in the Caribbean: Perceptions of Environmental Change. M.A. thesis. University of Waterloo. Waterloo, Canada.

Shiomi, K., M. Hosaka, S. Fujita, H. Yamanaka, and T. Kikuchi. 1989. Venoms from six species of marine fish: lethal and hymolytic activities and their neutralization by commercial stonefish antivenom. *Marine Biology* **103**: 285-289.

Smith, N. 2010. Lionfish invasion in nearshore waters of the Bahamas: an examination of the effects of artificial structures and invader versus native species colonization rates. M.Sc. thesis. University of British Columbia. Vancouver, Canada.

Sri Balasubashini, M., S. Karthigayan, S.T. Somasundaram, T. Balasubramanian, P. Viswanathan, and V.P. Menon. 2006a. In vivo and in vitro characterization of the biochemical and pathological changes induced by lionfish (*Pterios volitans)* venom in mice. *Toxicology Mechanisms and Methods* **16**: 525-531.

Sri Balasubashini, M., S. Karthigayan, S.T. Somasundaram, T. Balasubramanian, P. Viswanathan, P. Raveendran, and V.P. Menon. 2006b. Fish venom (*Pterois volitans*) peptide reduces tumor burden and ameliorates oxidative stress in Ehrlich's ascites carcinoma xenografted mice. *Bioorganic and Medicinal Chemistry Letters* **16**: 6219-6225.

Sullivan Sealy, K. et al. 2006. *Tools and Methods for Studies in Coastal Ecology of the Bahamas*. University of Miami, Coral Gables, Florida USA. 111 pp. Available: http://henge.bio.miami.edu/coastalecology/TOOLS%20&20METHODS%20April.pdf.

Vásquez-Yeomans, L., L. Carrillo, S.M.E. Estrella Malca, J.A. Morris, Jr., T. Schultz, J.T. Lamkin. 2011. First larval record of *Pterois volitans* (Pisces: Scorpaenidae) collected from the ichthyoplankton in the Atlantic. *Biological Invasions* **13**: 2635-2640.

Vetrano, S.J., J.B. Lebowitz, and S. Marcus. 2002. Lionfish envenomation. *Journal of Emergency Medicine* **23**: 379-382.

Washington, B.B., H.G. Moser, W.A. Laroche, and W.J. Richards. 1984. Scorpaeniformes: development. Pp. 438-447 in: Moser, H.G., J. Richards, D.M. Cohen, M.P. Fahay, A.W. Kendall, and S.L. Richards (eds.). *Ontogeny and Systematic of Fishes.* Special Publication 1 of the American Society of Ichthyologists and Herpetologists. Allen Press, Lawrence, Kansas USA.

Whitfield, P.E., J.A. Hare, A.W. David, S.L. Harter, R.C. Muñoz, and C.M. Addison. 2007. Abundance estimates of the Indo-Pacific lionfish *Pterois volitans/ miles* complex in the Western North Atlantic. *Biological Invasions* **9**: 53-64.

Wilkinson, C., A. Green, J. Almay, and S. Dionne. 2003. Monitoring Coral Reef Marine Protected Areas. Australian Institute of Marine Science and the IUCN Marine Program. Available: http://www.reefresilience.org/pdf/mcrmpa-v1.pdf.

Zuur, A.F., E.N. Leno, and C.S. Ephick. 2010. A protocol for data exploration to avoid common statistical problems. *Methods in Ecology and Evolution* **1**:3-14.

Zuur, A. F., E.N. Leno, N.J. Walker, A.A. Saveliev, and G. M. Smith. 2009. *Mixed Effects Models and Extensions in Ecology with R*. Statistics for Biology and Health Series, Springer Science+Business Media LLC., New York, New York USA. 547 pp.

Zuur, A.F., E.N. Leno, and G.M. Smith. 2007. *Analyzing Ecological Data*. Statistics for Biology and Health Series, Springer Science+Business Media LLC. New York, New York USA. 672 pp.